Letts
EDUCATIONAL

Exam Practice
A LEVEL

A Level
Exam Practice

Covers AS and A2

Physics

Author

Gurinder Chadha

Contents

AS and A2 exams

Different types of questions

In both AS and A2 examinations, there are mainly two sorts of examination questions: **structured** questions and **extended writing** questions.

Structured questions

In a structured question, a collection of common ideas is tested and the question is set out in smaller sub-sections. The earlier sections of the questions have a tendency to be easier and are designed to ease you into the question. Sometimes the answer to a particular section is given, so as to point you in the right direction. The later sections of a question can be a little discriminatory and you might have to think very carefully about your response. The following guide may be helpful:

- Read each question with great care and underline or highlight any key terms or data given.
- Write your answers carefully and always show **all** stages of your calculations.
- At every opportunity, keep referring to any introductory section to the question.

Extended questions

In an extended question, examiners will be assessing your understanding of physics and your ability to communicate ideas effectively and clearly. It is vital that in such questions, you pay particular attention to spellings, grammar and sentence construction. The response to such question can be open-ended, with examiners keen to award marks for any 'good', but relevant, physics. The following guide may be helpful:

- Read the question carefully and make sure that you do clearly understand what the examiners want.
- Make a quick and short plan of your key ideas. It is worth putting your ideas in a **hierarchical** order.
- Write your answers clearly and remember to pay particular attention to the way you write.

Other questions

In addition to structured and extended questions, examiners also make use of free-response and open-ended questions. The points raised in the guide above are equally applicable here.

What examiners look for

An examination paper is devised to assess your knowledge, understanding and application of physics. One of the main purposes of an examination paper is to sort out how good you are at physics. Examiners are normally looking for the following points:

- Correct answers to the questions.
- Clearly presented answers with all the working shown.
- Concise answers to structured questions and logically set out responses to extended questions.
- Sketch graphs and diagrams that are drawn neatly, with particular attention given to labels and units.

What makes an A, C and E candidate?

The examiners can only ask questions that are within the specifications (this is the new term for the 'syllabus'). It is therefore vital that you are fully aware of what the examiners can and cannot ask in any examination. Be prepared to tackle the complexities of the paper. Your aim must be to achieve high marks in unit or module examinations. The way to accomplish this is to have a good knowledge and understanding of physics. Listed below are the minimum marks for grades A to E.

      Grade A 80 %. **Grade B** 70 %. **Grade C** 60%. **Grade D** 50%. **Grade E** 40%.

- **A grade candidates** have an excellent all-round knowledge of physics and they can apply that knowledge to new situations. Such candidates tend to be strong in all of the modules and tend to have excellent recall skills.
- **C grade candidates** have a reasonable knowledge of physics but have some problems when apply their knowledge to new situations. They have some gaps in their knowledge and tend to be weak in some of the modules.
- **E grade candidates** have a poor knowledge of physics and have not learnt to apply their ideas to familiar and new situations. Such candidates find it difficult to recall keys definitions and equations.

Successful revision

Revision skills

- Always start with a topic that you find easier. This will boost your self-confidence.
- Do not revise for too long. When you are tired and irritable, you cannot produce quality work.
- Make notes on post cards or lined paper of key ideas and equations. Do not feel that you have to write down everything. Just the key points need to be jotted down. Sometimes you have to learn certain proofs. It is worth writing down all the important steps for such proofs.
- Make good use of the specification. Use a highlighter pen to identify topics that you have already revised.
- Do not leave your revision to the last moment. Plan out a strategy spread over many weeks before the actual examination. Work hard during the day and learn to relax when needed.
- Whatever happens, do not try to learn any new topics on the day before the examination. It is important for you to be calm and relaxed for the actual examination.

Practice questions

This book is designed to improve your understanding of physics and of course, improve your final grade.

Look carefully at the grade A and C candidates' responses. Can you do better? There are some important tips given to improve your understanding.

Try the practice examination questions and then look at the answers and tips given.

When you are ready, try the AS and A2 mock examination papers.

Planning and timing your answers

- Write legibly and stay focused throughout.
- Sometimes, candidates think that they have answered all the questions and then find an entire question on the last page. You do not want to be in this predicament, so **quickly** scan through the entire paper to see what you have to do.
- Do the question on the paper that you are most comfortable with. This will boost your confidence.
- Read each question carefully. Highlight the key ideas and data. The information given is there to be used.
- As a very rough guide, you have about one minute for each mark. The number of lines allocated for your answer gives you an idea of the depth and detail required for a particular answer. The marks allocated for each sub-section gives an idea of how many steps or items of information are required.
- Do a quick plan for extended questions. It is not sensible to start writing straight away because you will end up either repeating yourself or missing out some important ideas.
- Do not use correction fluid. If you have strong reasons that a particular answer is wrong, then simply cross it out and provide an alternative answer.

Setting out numerical answers

It is important that your answers to numerical and algebraic questions are set out logically for the examiner. In this book, a simple method is used to indicate **where** a mark is awarded for the correct response. This is indicated by means of a tick ($\checkmark$).

The **marking scheme** adopted in this book and how you ought to **structure** a numerical answer is illustrated in the example below.

Question: What is the pressure exerted by a force of 9.0 kN acting on an area of 1.5×10^{-2} m^2? **[2]**

Answers:

$P = F/A$ $\checkmark$ (Make the physics clear to the examiner.)

$P = 9.0 \times 10^3/1.5 \times 10^{-2}$ (Use standard form and remember to convert k $\rightarrow 10^3$.)

$P = 6.0 \times 10^5$ Pa $\checkmark$ (Do not forget the correct unit and significant figures.)

There are only two marks for the calculation. One mark is awarded for the equation and the other for the correct answer and the unit.

Remember, the ticks appear next to responses where the marks are awarded.

No credit can be given for a bold wrong answer. However, by writing down all the stages of your work, it may be possible to pick up some or all of the 'part marks'. So help yourself and set out your work in a clear and methodical way.

How to boost your grade

Boost your grade

Examiners cannot give credit for the wrong physics. A wrongly quoted equation cannot be awarded any marks. For an examiner, it is quite disturbing to find candidates that cannot re-arrange equations. Mathematics is the language of physics, therefore it is important for candidates to be comfortable with handling and re-arranging equations. There are several techniques for re-arranging equations, but the one outlined below can be learnt and applied quickly.

Remember BODMAS.

When solving a numerical or algebraic equation, you must do the mathematical operations in the order given by the mnemonic BODMAS.

$$\textbf{B}\text{racket} \rightarrow \textbf{O}\text{f} \rightarrow \textbf{D}\text{ivision} \rightarrow \textbf{M}\text{ultiplication} \rightarrow \textbf{A}\text{ddition} \rightarrow \textbf{S}\text{ubtraction}$$

When it comes to re-arranging an equation, you simply **reverse** the mathematical operations. Here is an example to illustrate this technique for re-arranging an equation.

$v^2 = u^2 + 2as$ What is a?

Using the ideas developed above, we have

$$a \rightarrow (\times 2s) \rightarrow (+ u^2) \rightarrow = v^2$$

By reversing the sequence and carrying out the inverse operations, we end up with

$$v^2 \rightarrow (- u^2) \rightarrow (\div 2s) \rightarrow = a$$

Therefore $a = \dfrac{(v^2 - u^2)}{2s}$

If you have some other tried and tested technique for re-arranging equations, then it is best to stick to it. However, do remember to take re-arranging of equations seriously in physics.

Here are some other suggestions to **boost your grade**.

- Be familiar with the specifications.

- Learn all the definitions within the specifications. Recalling definitions can give you easy marks and improve your final grade.

- Write all the stages of a numerical solution. If your final answer is wrong, you still have chance to pick up some of the 'part-marks'.

- In a question with 'state', the answer is brief and does not require any further explanation. In a question with 'describe', the answer can be long and may require full explanation of some physics.

- Use the information given in the question to guide your answers. For numerical solutions, keep an eye on the significant figures and units. Your final answer must not be more or less the significant figures given in the question. In physics, it is sensible to write the final answer in standard form, e.g.: 1.62×10^{-3} A.

- Read information given on graphs and tables carefully. Sometimes data is given in either standard or prefix form. Do not forget to take this on board when doing your calculations. For example, the stress axis is labelled as 'stress/MPa'. Remember that 'MPa' is 10^6 Pa.

- It is easy to press the wrong buttons on the calculator. Make sure that your answer looks reasonable. If you have time, you ought to check a calculation again.

- Draw diagrams carefully and make sure that you label all the key items.

- Your graphs must be correctly labelled and have a suitable scale so that it fills most of the graph paper.

- You do not have to recall physical data. All the data required is normally given on the question paper itself or on a separate data-sheet.

Questions with model answers

C grade candidate – mark scored 6/10

 For help see Revise AS Study Guide pages 31 and 32

(1) Explain what is meant by an **elastic** material. **[1]**

> *A material that returns to its original shape and size when the forces are removed.* ✔

(2) With the aid of a sketch graph, show the force–extension graph for a typical ductile material. **[2]**

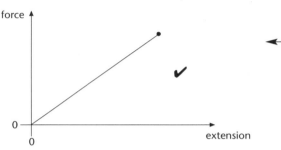

✔

Examiner's Commentary

There is no evidence of what happens to the material beyond the elastic limit, hence the candidate lost a valuable 'detail' mark. If the candidate had shown plastic deformation, then the second mark could have been scored.

(3) One of many cables supporting a small suspension bridge has a radius of 1.5 cm and a natural length of 9.0 m. The tension in each cable is 65 kN. The Young's modulus of the cable material is 2.1×10^{11} Pa.

(a) Show that the cable has a cross-sectional area of 7.1×10^{-4} m². **[1]**

> $area = \pi r^2$
> $area = \pi \times (1.5)^2 = 7.07$ ✗

The candidate did not convert the radius into metres, the answer is therefore out by a factor of 10^4.

(b) Calculate the stress in each cable. **[2]**

> $stress = \dfrac{F}{A}$
> $stress = \dfrac{65 \times 10^3}{7.1 \times 10^{-4}}$ ✔
> $stress = 9.2 \times 10^7$ Pa ✔

Fortunately, the candidate used the information given in (a) and that was a sensible strategy.

(c) Calculate the strain experienced by the cable and its extension. State any assumption made. **[4]**

> $E = \dfrac{stress}{strain}$
> $strain = \dfrac{stress}{E}$
> $strain = \dfrac{9.2 \times 10^7}{2.1 \times 10^{11}}$
> $strain = 4.4 \times 10^{-4}$ ✔
> *The material obeys Hooke's law.* ✔

The question also required the extension of the cable, this was not done by the candidate. Since
$strain = \dfrac{extension}{original\ length}$
the extension can be calculated from the value of the strain. The correct value for the extension is 3.9×10^{-3} m.

Questions with model answers

A grade candidate – mark scored 6/7

 For help see Revise AS Study Guide pages 28, 48, 59 and 63

The diagram shows a person pushing a roller at a constant velocity of 0.9 ms^{-1} along a flat horizontal ground. The handle makes an angle of 37° to the horizontal and a force of 60 N is directed along the handle as shown on the diagram.

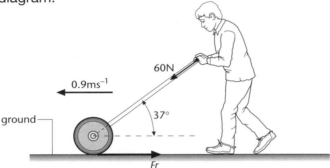

Examiner's Commentary

(1) Calculate the horizontal component of the force acting on the roller. **[2]**

$F_x = F\cos\theta$ ✔
$F_x = 60 \times \cos 37 = 48$ N ✔

*The roller is moving at a constant velocity and therefore has no acceleration. According to $F = ma$, there is no **net** force acting on the roller. The candidate's answer is wrong because there are several forces acting on the roller. It so happens that horizontally, there is no **resultant** force.*

(2) What is the magnitude of the frictional force F_r on the roller? Explain your answer. **[2]**

Friction F_r is equal to 48 N ✔
because there is no force on the roller. ✗

(3) Calculate the rate of work done in pushing the roller along the ground. **[3]**

work = Fx
In one second, the distance moved is 0.9 m.
Work done in 1s is the power. Therefore
power = 48 × 0.9 ✔
power = 43 W ✔ ✔

One mark was reserved for the correct unit for the rate of working or the power.

Exam practice questions

A *Answers on pp. 13–16*

(1) **(a)** Distinguish between a *scalar* quantity and a *vector* quantity. **[2]**

(b) A car travels one complete lap around a circular track at an average speed of 100 km hr^{-1}.

(i) If the lap takes 3.0 minutes, show that the length of the track is 5.0 km. **[2]**

(ii) What is the magnitude of the displacement of the car after 1.5 minutes? **[2]**

[AQA Specimen, 2001/2]

(2) **(a)** State the principle of moments. **[2]**

(b) The diagram shows a **model** for a human arm that is balanced in the horizontal position.

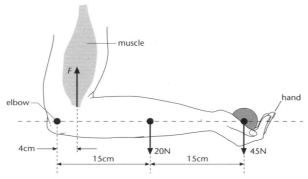

The arm is assumed to be uniform and its centre-of-gravity is 15 cm away from the elbow. The weight of the arm is 20 N and the hand is holding an object of weight 45 N.

(i) Explain what is meant by *centre-of-gravity*. **[1]**

(ii) Calculate the vertical force *F* provided by the arm muscle. **[3]**

(iii) The arm is extended further away from the body but still balanced in the horizontal position. The arm muscle now exerts a force *F* at an angle to the vertical. This is shown in the diagram below.

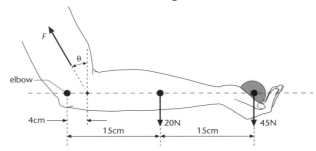

Without any further calculations, explain how the magnitude of the force *F* changes as compared with your answer to **(b) (ii)**. **[2]**

Exam practice questions

(3) In order to display greeting cards, a student fixes a length of string between two nails and then suspends the cards from the string. The diagram shows the string with one card of weight 0.60 N suspended by a light clip at the centre of the string.

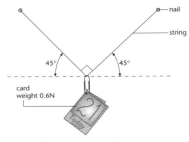

(a) On the diagram, mark the forces on the clip due to the tension in the string.   **[2]**

(b) The resultant of the forces due to the tension in the string is 0.60 N.
In the space below, draw a vector triangle for the forces in the string and their resultant.

                                                                                        [3]

(c) Use the completed vector diagram to determine the magnitude of the tension in the string.

tension = ... N   **[1]**

[OCR Specimen, 2000]

(4) **(a)** State Hooke's law.   **[2]**

(b) The area under a force–extension graph is equal to work done by the force. Use this idea and the sketch graph below to find an expression for the energy stored in a spring in terms of the applied force F and the final extension e.

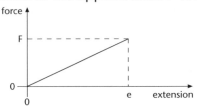

                                                                                        [2]

(c) A diagram of a child's toy is shown below.

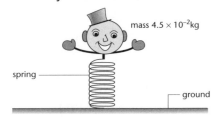

The total mass of the toy and spring is 4.5×10^{-2} kg. A force of 12 N compresses the spring by 5.0 cm.

(i) Calculate the energy stored in the spring when it is compressed by 5.0 cm. [2]

(ii) When the spring is released, the toy lifts off the ground and reaches a maximum vertical height h above the ground. Calculate this height h. State any assumption made.
Data: $g = 9.8 \text{ Nkg}^{-1}$ [3]

(5) **(a)** Define acceleration. [1]

(b) A 1200 kg car is travelling towards a rigid safety barrier at a velocity of 28 ms^{-1}. The diagram shows the velocity–time graph for the car when colliding with this barrier.

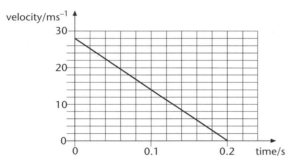

(i) Describe the motion of the car as it hits the barrier. [2]

(ii) Calculate the magnitude of the force exerted by the barrier on the car. [3]

(iii) Safety barriers and cars are designed to **crumple** on impact. What effect does this have on car safety? [2]

(6) In an accident on a motorway, a car of mass 950 kg leaves a skid mark 20 m long when stopping. The accident investigators suspect the deceleration of the car to be 12 ms^{-2}.

(a) Calculate the magnitude of the average braking force between each of the four car tyres and the road. [2]

(b) Calculate the initial speed of the car. [2]

(7) **(a)** Write an equation for the kinetic energy of an object. Define any symbols used. [2]

Exam practice questions

(b) A 3.2×10^{-2} kg metal ball is **dropped** from a vertical height 2.0 m. It hits the ground and makes a dent of depth 4.0 mm.

    (i) Show that the impact velocity of the metal ball with the ground is
        6.3 ms^{-1}. You may assume that there is negligible air resistance.
        Data: g = 9.8 ms^{-2} **[2]**

   (ii) Calculate the kinetic energy of the metal ball just before it
        hits the ground.   **[2]**

  (iii) Calculate the mean deceleration of the ball during its impact
        with the ground.   **[2]**

(8) The diagram shows an 8.0 kg shopping bag placed on the floor of a lift.

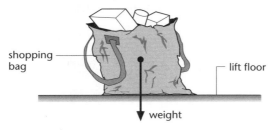

    shopping bag   lift floor

            weight

(a) Complete the diagram to show the contact force provided by the
floor on the bag. Label the force R.   **[1]**

(b) Calculate the weight of the shopping bag.
Data: g = 9.8 N kg^{-1} **[1]**

(c) Calculate the contact force R provided by the floor when the lift is moving
vertically upwards

    (i) at a constant velocity of 0.50 ms^{-1},   **[1]**

   (ii) at an acceleration of 2.5 ms^{-2}.   **[2]**

Answers

(1) (a) A scalar quantity has only size (or magnitude).
A vector quantity has both magnitude **and** direction.

(b)(i) $v = \dfrac{x}{t}$

$x = vt = \dfrac{100 \times 3}{60}$

$x = 5\text{ km}$

Examiner's tip

You can work in S.I. units. This route, however, involves some conversions. The speed v is

$v = \dfrac{100 \times 10^3}{3600} = 27.78\text{ ms}^{-1}$

and the time t is

$t = 3.0 \times 60 = 180s.$

Therefore,

$x = 27.78 \times 180$

$x = 5.0 \times 10^3\text{ m.}$

(ii) After 1.5 mins, the car will have a displacement equal to the diameter of the track.

circumference = πd

displacement = $\dfrac{5.0}{\pi}$

displacement = 1.59 km

Examiner's tip

There is one mark for the correct unit.
*In this part of the question, you need to be clear about **displacement** and **distance**.*
The distance covered by the car is equal to half the circumference of the track and this is <u>not</u> what is required in the question.

(2) (a) For **rotational** equilibrium,
the sum of the clockwise moments = sum of anticlockwise moments.
The moments are taken about the same point or pivot.

(b)(i) The centre-of-gravity is the point at which the weight of the object appears to act.

(ii) Take moments about the elbow.
sum of anticlockwise moments = sum of clockwise moments

$4 \times F = (15 \times 20) + (45 \times 30)$

$F = \dfrac{1650}{4} = 413 \approx 410\text{ N}$

Examiner's tip

*You can work in metres. If you do, the 10^{-2} factor on both sides of the equation cancels out. Do remember that each moment is taken about the elbow, so all distances must be measured from **this** point. The downward force of 45 N is 30 cm from the elbow and not 15 cm.*

Answers

(iii) The force *F* is greater than before.
The anticlockwise moment must still be the same. Hence, the **vertical**
component of the force (given by $F\cos\theta$) must be 410 N as in **(b)(ii)**.
This implies that the force *F* exerted by the muscle must be greater
than before.

(3) (a) There is a force vertically downwards equal to
0.60 N due to the weight of the card.
The tension in both strings provides the necessary upward force.

(b) Correct shape of triangle.
Forces shown on the diagram.
Correct direction of forces.

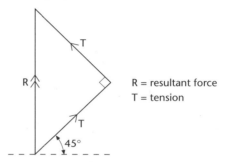

R = resultant force
T = tension

Examiner's tip

*The tensions in both strings add together to provide a resultant force which must be
0.60 N in magnitude. The direction of the resultant force must be opposite to that of the
weight of the card.*

(c) Using Pythagoras' theorem $\Rightarrow$
$T^2 + T^2 = 0.60^2$
$2T^2 = 0.036$
$T \approx 0.42$ N

(4) (a) Extension of material is directly proportional to the applied force.
This is true as long as the elastic limit (or limit of proportionality) is not exceeded.

(b) Area $= \frac{1}{2}bh$
energy $= \frac{1}{2}Fe$

Examiner's tip

*There are other assumptions as
well. The presence of air resistance
will reduce the height attained by
the toy. Try not to be vague with
your answer. An answer like 'the
spring is not 100% efficient' is
unclear in its physics.*

(c)(i) energy $= \frac{1}{2}Fe$
energy $= \frac{1}{2} \times 12 \times 5.0 \times 10^{-2}$
energy $= 0.30$ J

(ii) Assumption: gravitational P.E. =
energy stored in spring
$mg\Delta h = 0.30$
$h = \dfrac{0.30}{4.5 \times 10^{-2} \times 9.8}$
$h = 0.680 \approx 0.68$ m

Examiner's tip

*The gradient from a velocity–time
graph is equal to acceleration. The
gradient of the line is constant but
negative. Therefore, the car's
velocity is decreasing at a
constant rate.*

(5) (a) acceleration =
rate of change of velocity

(b)(i) The car experiences a **constant**
deceleration.

(ii) acceleration = gradient

$$a = \frac{-28}{0.2} = -140 \text{ ms}^{-2}$$

$$F = ma$$

$$F = 1200 \times 140 = 1.68 \times 10^5 \approx 1.7 \times 10^5 \text{ N}$$

Examiner's tip

The magnitude of the force is required, so the negative sign in the answer is optional.

(iii) Impact forces are smaller
because the time of impact is longer (hence smaller deceleration).

(6) (a) $F = ma$

$$F = 950 \times 12 \qquad \text{(magnitude only)}$$

$$F = 1.14 \times 10^4 \text{ N}$$

Force on each tyre $= \dfrac{1.14 \times 10^4}{4}$

Force on each tyre $= 2.85 \times 10^3 \approx 2.9 \times 10^3 \text{ N}$

(b) $v^2 = u^2 + 2as$

$$a = -12 \text{ ms}^{-2} \qquad v = 0 \qquad s = 20\text{m} \qquad\qquad u = \text{initial velocity}$$

$$u^2 = 2 \times 12 \times 20$$

$$u = 21.9 \approx 22 \text{ ms}^{-1}$$

Examiner's tip

Instead of kinematics, you can solve the problem using energy considerations.

Work done to stop car = Initial K.E. of car

$$Fs = \tfrac{1}{2} m u^2$$

$$u = \sqrt{\frac{2Fs}{m}}$$

$$u = \sqrt{\frac{2 \times 1.14 \times 10^4 \times 20}{950}}$$

$$u \approx 22 \text{ ms}^{-1}.$$

(7) (a) K.E $= \tfrac{1}{2} mv^2$

m is the mass of object and *v* is its speed.

(b)(i) $v^2 = u^2 + 2as$

$u = 0$, therefore $v^2 = 2as$

$$v = \sqrt{2as} = \sqrt{2 \times 9.8 \times 2.0}$$

$$v = 6.26 \text{ ms}^{-1}$$

Examiner's tip

Remember you cannot score any marks if no analysis is shown since the answer is already given in the question. So you cannot provide a bare ' v = 6.3 ms⁻¹ ' as your answer. You can also get the answer by using the principle of conservation of energy. The gravitational potential energy of the ball at the start will be equal to its kinetic energy at the ground. Therefore

$$mg\Delta h = \tfrac{1}{2} mv^2$$

$$3.2 \times 10^{-2} \times 9.8 \times 2.0 = \tfrac{1}{2} \times 3.2 \times 10^{-2} \times v^2$$

Hence v = 6.26 ms⁻¹.

Answers

(ii) K.E. $= \frac{1}{2} mv^2$

K.E. $= \frac{1}{2} \times 3.2 \times 10^{-2} \times 6.26^2$

K.E. $= 0.627 \approx 0.63$ J

Examiner's tip

The answer for the speed is provided in (b)(i) above. You can take advantage of this if you have been unsuccessful with calculating the speed of impact.

(iii) Work done = K.E. of the ball

$Fs = 0.627$

(F = force on the ball)

$F = \dfrac{0.627}{4.0 \times 10^{-3}}$

$F = 1.57 \times 10^2 \approx 1.6 \times 10^2$ N

$a = \dfrac{F}{m}$

$a = \dfrac{1.57 \times 10^2}{3.2 \times 10^{-2}}$

$a = 4.91 \times 10^3 \approx 4.9 \times 10^3$ ms^{-2}

Examiner's tip

The size of the dent made by the ball is given in millimetres, it is vital that you do convert the distance given into metres. An alternative route to the correct answer would be to use the equation

$$v^2 = u^2 + 2as.$$

(8) (a) The contact force is vertically upwards.

(b) weight $= mg$

weight $= 8.0 \times 9.8 = 78.4 \approx 78$ N

(c)(i) $F = ma$

Since $a = 0$, the **net** force F on the bag $= 0$

Therefore $R =$ weight

$R \approx 78$ N

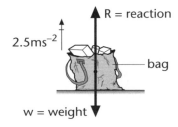

R = reaction

2.5ms^{-2}

bag

w = weight

(ii) Net force vertically $= ma$

$R - W = 8.0 \times 2.5$

$R - 78.4 = 20$

$R = 98.4 \approx 98$ N

Examiner's tip

It is very tempting to just use 'R = ma'. This is, of course, not the case. The force R must be greater than the downward force W in order to provide the bag with an upward acceleration.

Questions with model answers

C grade candidate – mark scored 6/10

 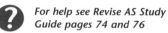 *For help see Revise AS Study Guide pages 74 and 76*

Examiner's Commentary

(1) Complete the sentence below.

Electric current is equal to the rate of flow of *charge* ✔ **[1]**

(2) Show that the alternative unit for current is C s⁻¹. **[1]**

$I = \dfrac{\Delta Q}{\Delta t}$ $I \rightarrow$ *[coulombs / seconds]* ✔

Hence, current $\rightarrow [C\,s^{-1}]$

The answer is correct. A shorter but alternative route would be to use $P = \dfrac{V^2}{R}$.

$R = \dfrac{V^2}{P}$

(3) A washing machine, rated as '240 V, 1 kW', is operated for 30 minutes.

   (a) Calculate the resistance of the appliance. **[2]**

$P = VI$ $I = \dfrac{P}{V} = \dfrac{1000}{240}$ $\therefore I = 4.17\,A$ ✔

$V = IR$ (*Ohm's law*)

$R = \dfrac{V}{I} = \dfrac{240}{4.17}$ $R = 58\,\Omega$ ✔

$\therefore R = \dfrac{240^2}{1000}$

$R = 58\,\Omega$

   (b) Calculate the charge flow through the appliance in a period of 30 minutes. Hence determine the number of electrons responsible for this flow of charge.
Data: $e = 1.6 \times 10^{-19}$ C. **[3]**

$\Delta Q = I\,\Delta t$

$\Delta Q = 4.17 \times (30 \times 60) = 7.5 \times 10^3\,C$ ✔

number of electrons $= 7.5 \times 10^3 \times 1.6 \times 10^{-19}$ ✗

number $= 1.2 \times 10^{-15}$ ✗

The candidate managed to secure one mark for calculating the charge flow in the time of 30 minutes. It was sensible to convert the time into seconds. The total charge of 7.5×10^3 C is due to N number of electrons, each carrying a charge of 1.6×10^{-19} C. Therefore:

$7.5 \times 10^3 = N\,e$

$N = \dfrac{7.5 \times 10^3}{1.6 \times 10^{-19}}$

$N = 4.7 \times 10^{22}$

   (c) What is the cost of using the appliance for 30 minutes? The cost of each kW h of energy is 6.4 p. **[3]**

$E = P\Delta t$ $E = 1000 \times (30 \times 60) = 1.8 \times 10^6\,J$ ✔

$1\,kW\,h = 1 \times 3600 = 3600\,J$ ✗

$\therefore cost = \dfrac{3600 \times 6.4}{1.8 \times 10^6}$

$= 0.013\,p$ ✗

The amount of energy transformed by the appliance is correct. The number of 'units' of energy transformed is given by

number of 'units' =
$1\,kW \times (30/60)\,h$
$= 0.50\,kW\,h$
$\therefore cost =$
$0.50 \times 6.4 = 3.2\,p$

A grade candidate – mark scored 8/10

The diagram shows an electrical circuit based on two switches A and B. The supply may be assumed to have negligible internal resistance.

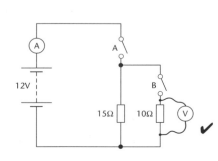

✔

Questions with model answers

A grade candidate continued

 For help see Revise AS Study Guide pages 71, 76 and 77

(1) On the diagram, show how the potential difference across the 10 Ω resistor may be measured. **[1]**

(2) Calculate the ammeter reading when A is **closed** and B is **open**. **[2]**

Current is only in the 15 Ω resistor.

$$I = \frac{V}{R} = \frac{12}{15} \quad ✔$$

$$I = 0.8 \ A \quad ✔$$

(3) A and B are **both** closed.

(a) Calculate the new ammeter reading. **[3]**

$$\frac{1}{R} = \frac{1}{15} + \frac{1}{10} \quad ✔$$

$$\frac{1}{R} = \frac{25}{150} = 0.167$$

$$R = \frac{1}{0.167} = 6.0 \ \Omega \quad ✔$$

$$I = \frac{V}{R} = \frac{12}{6.0}$$

$$I = 2.0 \ A \quad ✔$$

(b) Calculate the ratio

$$\frac{\text{power dissipated in 15 } \Omega \text{ resistor}}{\text{power dissipated in 10 } \Omega \text{ resistor}}. \quad \textbf{[2]}$$

$$P = \frac{V^2}{R}$$

For 10 Ω resistor: $P = \dfrac{12^2}{10} = 14.4 \ W$

For 15 Ω resistor: $P = \dfrac{12^2}{15} = 9.6 \ W \quad ✔$

$$\therefore \text{Ratio} = \frac{14.4}{9.6} = 1.5 \quad ✗$$

(c) State and explain if the ratio calculated in **(b)** would change, if at all, when a 6.0 V supply of negligible internal resistance is used. **[2]**

The ratio will not change ✔

because the resistors are connected in parallel. ✗

Examiner's Commentary

With the switch B open, there can be no current in the 10 Ω resistor. Before substituting any numbers into the V = IR equation, it is worth spending a few moments scrutinising the circuit.

Under the time restrictions of an exam, candidates often forget to inverse their answer when using $\dfrac{1}{R} = \dfrac{1}{R_1} + \dfrac{1}{R_2}$. *To avoid this, another equivalent equation for determining the total resistance may be used. For* **two** *resistors of resistance values R_1 and R_2 connected in parallel, the total resistance R_T, is given by*

$$R_T = \frac{R_1 R_2}{(R_1 + R_2)}$$

Therefore: $R_T = \dfrac{15 \times 10}{(15 + 10)}$

$R_T = 6.0 \ \Omega$

After all the hard work, the candidate calculated the ratio incorrectly. When calculating the electrical power, the candidate has used 12 V for both resistors. Since the p.d. across a parallel circuit is the same, it follows that $P \propto \dfrac{1}{R}$. *The ratio will therefore be*

$$\text{ratio} = \frac{10}{15}$$

$\text{ratio} = 0.67$

The ratio will be the same as before, and for this, the candidate has scored the mark. The reason for the ratio being the same lacks clarity. A response like: 'The p.d. across each resistor is the same and because $P \propto \dfrac{1}{R}$. The ratio only depends on the values of the resistors.' could score both marks.

Exam practice questions

A *Answers on pp. 23–27*

(1) **(a)** Explain what is meant by electric current. **[1]**

(b) A student wishes to determine the resistance of an electrical component.
  (i) Name two quantities required to determine resistance. **[2]**

  (ii) In the space provided, complete the circuit diagram to show how appropriate meters may be used to determine the resistance of a resistor.

[2]

(c) **(i)** State Ohm's law. **[2]**

  (ii) Name a component that does **not** obey Ohm's law. **[1]**

  (iii) **1.** Sketch the current-voltage characteristic of a semiconducting diode.

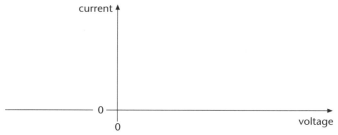

[2]

  2. Explain how the resistance of the semiconducting diode changes as the voltage (potential difference) is increased from negative to positive values. **[3]**

(2) **(a)** The current *I* in a metallic wire is given by the equation
$$I = Anev$$
Define the symbols used in this equation. **[2]**

(b) A metallic material is connected to a battery. The cross-sectional shape of the material is shown below.

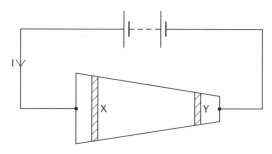

Exam practice questions

    (i) Suggest why the current at sections X and Y is the **same**.   **[1]**

    (ii) State and explain what happens to the mean drift velocity of the electrons at Y compared with that at X.   **[2]**

    (iii) Explain why the end Y is warmer than the end X.   **[2]**

(3) A light-dependent resistor may be used with additional components to make a light-meter. Sketch a diagram of a suitable circuit.   **[2]**

Explain how the circuit works.   **[2]**

[Edexcel Specimen, 2000]

(4) A battery of e.m.f. 12 V and internal resistance r is connected in a circuit with three resistors, each having a resistance of 10 Ω as shown. A current of 0.50 A flows through the battery.

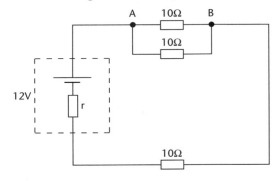

Calculate
(a) the potential difference between A and B in the circuit,   **[2]**

(b) the internal resistance of the battery,   **[2]**

(c) the total energy supplied by the battery in 2.0 s,   **[1]**

(d) the fraction of energy supplied by the battery and that which is dissipated within the battery.   **[2]**

[AQA, 2000]

(5) **(a)** Define electrical resistance.   **[1]**

(b) Show that electrical resistivity has the unit Ωm.   **[2]**

(c) A car lamp is rated as '12 V, 60 W'. The lamp has a filament wire of radius 3.2×10^{-4} m and when operating at 12 V, the metal of the filament has resistivity 4.2×10^{-7} Ωm.

    (i) Calculate the resistance of the lamp when operated at 12 V.   **[2]**

    (ii) Calculate the length of the filament wire and comment on its value.   **[3]**

(6) The diagram shows a potential divider circuit based on a light-dependent resistor (LDR). Assume the supply has negligible internal resistance.

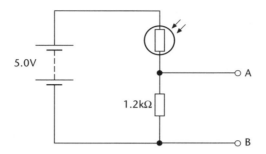

(a) State how the resistance of the LDR is affected by the intensity of light falling on it. **[1]**

(b) At a particular intensity of light, the LDR has a resistance of 820 Ω. Calculate the potential difference between A and B. **[3]**

(c) On the axes provided, draw a sketch graph to show the variation of the potential difference V between A and B with light intensity.

[2]

(7) **(a)** Explain what is meant by an alternating current. **[1]**

(b) (i) Write an equation for the root-mean-square voltage, $V_{r.m.s}$ in terms of the peak voltage, V_o. **[1]**

(ii) The output signal from an electrical circuit is described as being sinusoidal and as having 'a frequency of 250 Hz and a peak voltage of 4.0 V'.

1. Calculate the period of the signal. **[2]**

2. Calculate the r.m.s. value of the voltage. **[2]**

Exam practice questions

3. The diagram shows the screen of an oscilloscope.

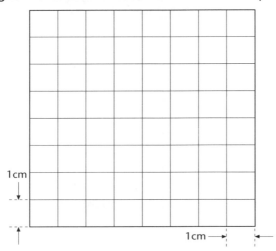

The oscilloscope is connected to the output of the electrical circuit. On the diagram above, show the appearance of the trace for this sinusoidal signal.

The oscilloscope sensitivity is 2.0 V/cm and its time base is set to 1.0 ms/cm.

[2]

Answers

(1) (a)   Electric current is the flow of charge.

(b)(i)   Voltage (or potential difference)
and current.

(ii)

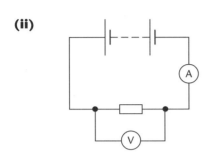

Ammeter connected in series.
Voltmeter connected across the resistor.

(c)(i)   The potential difference is directly proportional to the current
in a metallic conductor at constant temperature.

(ii)   Filament lamp or semiconducting diode.

(iii) 1.

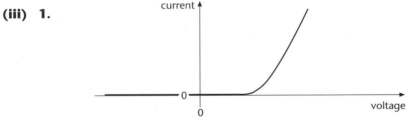

No current shown for negative values of voltage.
An increasing current shown for positive values of voltage.

Answers

2. For negative voltages, there is no current because
the diode has infinite resistance.
For positive voltages, the resistance of the diode
decreases as the voltage across it increases.
For silicon diode, the resistance is infinite for voltage
less than about 0.6 V.

(2) (a) $A \rightarrow$ <u>cross-sectional</u> area of the wire.
$n \rightarrow$ number density of charges or number of charge carriers per unit volume.
$e \rightarrow$ electronic charge.
$v \rightarrow$ mean drift velocity of the charges.

(b)(i)   The current is the same because
both sections X and Y are connected in **series**.

(ii)   The mean drift velocity at Y is greater than that at X.
$I = Anev$
Since I, n and e are the same for the material, we have
$v \propto \dfrac{1}{A}$.
The section Y has a smaller cross-sectional area and
therefore the mean drift velocity at Y is greater.

(iii)   The section Y is warmer than X because its resistance (per unit length)
is larger.
Power dissipated is given by $P = I^2 R$.
Since current I is the same, $P \propto R$.
Therefore section Y is warmer than X.

(3)

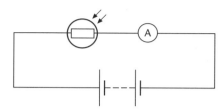

LDR connected in series with the battery and the ammeter.
Correct symbol used for the LDR.

As the intensity of incident light increases, the resistance of
the LDR decreases.
This leads to a greater current in the circuit because $I \propto \dfrac{1}{R}$

for a given supply.

Examiner's tip

*There is an alternative in the form of a potential divider circuit. The LDR can be
connected in series to the battery via a resistor. A potential difference across the resistor
can then be measured using a voltmeter.*

(4) (a) $R_T = \dfrac{R_1 R_2}{(R_1 + R_2)}$

$R_T = \dfrac{10 \times 10}{10 + 10} = 5.0 \ \Omega$

$V = IR$

$V = 0.50 \times 5.0 = 2.5 \ V$

(b) For the 10 Ω resistor $\Rightarrow$
$V = IR = 0.50 \times 10 = 5.0 \ V$
P.d. across the internal resistor,
$V_r = 12 - (5.0 + 2.5)$
$V_r = 4.5 \ V$
Therefore, $r = \dfrac{V}{I} = \dfrac{4.5}{0.50} = 9.0 \ \Omega$

Examiner's tip

*According to Kirchhoff's second law, the e.m.f.
is equal to the sum of the p.d.s. in a loop.*
$$E = \Sigma V \qquad E = V + Ir$$
$$\therefore r = \dfrac{(E - V)}{I}$$
$$r = \dfrac{4.5}{0.50} \qquad r = 9.0 \ \Omega$$

(c) energy = $P\Delta t$ where $P = EI$ E = e.m.f.
energy = $EI\Delta t$
energy = $12 \times 0.50 \times 2.0$
energy = 12 J

(d) energy supplied to internal resistor = $VI\Delta t = 4.5 \times 0.50 \times 2.0 = 4.5$ J
$$\dfrac{\text{energy supplied by battery}}{\text{energy supplied to internal resistor}} = f$$

$f = \dfrac{12}{4.5} = 2.67 \approx 2.7$

Examiner's tip

*Throughout this question, the most fundamental idea is that the current in a series
circuit remains the **same**.*

Answers

(5) (a) $\text{resistance} = \dfrac{\text{voltage}}{\text{current}}$

(b) $R = \dfrac{\rho\ell}{A}$

$\rho = \dfrac{RA}{\ell}$

$\therefore \rho \rightarrow [\Omega \times m^2/m] \rightarrow [\Omega\, m]$

(c)(i) $P = \dfrac{V^2}{R}$

$R = \dfrac{V^2}{P}$

$R = \dfrac{12^2}{60}$

$R = 2.4\ \Omega$

(ii) $A = \pi r^2 = \pi (3.2 \times 10^{-4})^2 = 3.217 \times 10^{-7}\ m^2$

$\ell = \dfrac{RA}{\rho}$

$\ell = \dfrac{2.4 \times 3.217 \times 10^{-7}}{4.2 \times 10^{-7}}$

$\ell = 1.84 \approx 1.8\ m$

Too long for the size of the bulb, therefore the wire must be coiled.

(6) (a) The resistance decreases as the incident intensity of light increases.

(b) $R_T = 1200 + 820 = 2020\ \Omega$

$I = \dfrac{V}{R} = \dfrac{5.0}{2020} = 2.475 \times 10^{-3}\ A$

$V = IR$

$V = 2.475 \times 10^{-3} \times 1200$

$V = 2.97 \approx 3.0\ V$

Examiner's tip

The procedure above is simple and shows a good understanding of a series circuit. The potential difference, V across a resistor is also given by

$$V = \dfrac{R_2}{(R_1 + R_2)} \times V_0$$

where V_0 is the total p.d. across the potential divider circuit and R_1 and R_2 are the resistances of the resistors. The output is taken across the resistor with resistance R_2. This equation gives

$$V = \dfrac{5.0 \times 1200}{(1200 + 820)}$$

$$V = 2.97 \approx 3.0\ V$$

(c) For no light, the resistance of LDR is very large, therefore the p.d. across 1.2 kΩ resistor is very small (almost zero). Curve 'tending to' $V \approx 5.0$ V in brighter light conditions.

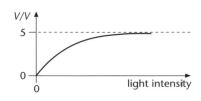

(7) **(a)** A current that changes direction.

(b)(i) $V_{r.m.s} = \dfrac{V_0}{\sqrt{2}}$

   (ii) **1.** $f = \dfrac{1}{T}$

$$T = \dfrac{1}{250}$$

$$T = 4.0 \times 10^{-3} \text{ s}$$

2. $V_{r.m.s} = \dfrac{4.0}{\sqrt{2}}$

$V_{r.m.s} = 2.83 \approx 2.8 \text{ V}$

3.

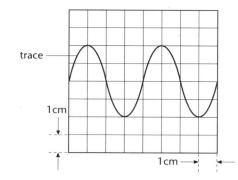

Trace showing one cycle completed in 4 cm on the screen.
Trace showing a signal with peak-to-peak voltage of 4 cm on the screen.

Questions with model answers

C grade candidate – mark scored 6/10

For help see Revise AS Study Guide pages 99, 100 and 101

In a data book, the following information is given about lithium.

Symbol: Li
Atomic (proton) number: 3
Isotopes: Lithium-6 (stable)
Lithium-7 (stable)
Lithium-8 (emits β particles)

(1) What are isotopes?   [2]

> *Isotopes are nuclei with the same number of protons* ✔
> *but different number of neutrons.* ✔

(2) For the lithium-7 isotope, state the number of

(a)   protons:

> *The nucleus has three protons.* ✔

(b)   neutrons: [2]

> $N = A - Z$
> $N = 7 - 3 = 4$ *There are four neutrons.* ✔

(3) The lithium-8 isotope emits beta (β) particles.

(a)   State the nature of beta particles.   [2]

> *β particles carry negative charge.* ✔

(b)   With reference to a nuclear decay equation, state and explain the change taking place within the nucleus of lithium-8 when it emits a beta particle.   [4]

> *The β particle carries energy away from the nucleus making it more stable.* ✔

Examiner's Commentary

The candidate is correct to refer to nuclei and not to atoms. An atom consists of a positive nucleus, but it also consists of the negatively charged electrons.

There are two marks and the candidate has given just one response. It would have been advantageous to state that 'beta particles are electrons that carry negative charge away from the nucleus.'

This is a very brief answer. The candidate was awarded one mark for appreciating that the decay is an attempt by the nucleus to become more stable. However, there is no nuclear decay equation. If this were to be done, then it would have been a prompt for the candidate to think in terms of the nucleon and the proton numbers. The decay equation for the nucleus is
$$^{8}_{3}Li \rightarrow ^{0}_{-1}e + ^{8}_{4}Be.$$
*The nucleon number remains the same, but there is one **extra** proton within the nucleus. This could only have happened if a neutron inside the nucleus transforms into a proton and an electron. The electron is the beta particle that is emitted by the unstable nucleus of lithium-8. Within the nucleus, the following change takes place*
neutron → proton + electron + anti-neutrino
$$^{1}_{0}n \rightarrow ^{1}_{1}p + ^{0}_{-1}e + \bar{\nu}_e$$

A grade candidate – mark scored 6/6

For help see Revise AS Study Guide pages 91 and 92

(1) Write the equation of state for an ideal gas. **[1]**

$PV = nRT$ ✔

(2) The diagram shows a metal cylinder in which some gas is trapped by a piston.

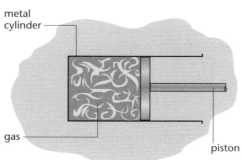

metal cylinder

gas

piston

(a) The volume of the trapped gas is 1.7×10^{-3} m³.
At a temperature of 210°C, the pressure exerted by the gas is 3.5×10^5 Pa. Calculate the number of moles of gas within the cylinder. You may assume that the trapped gas behaves like an ideal gas.
Data: R = 8.31 J mol⁻¹ K⁻¹ **[3]**

$$n = \frac{PV}{RT}$$

$T = 273 + 210 = 483\ K$ ✔

$$n = \frac{3.5 \times 10^5 \times 1.7 \times 10^{-3}}{8.31 \times 483}$$ ✔

$n = 0.15\ mol$ ✔

(b) The gas within the cylinder expands. The volume occupied by the gas **increases** by 30%, but the pressure is maintained at 3.5×10^5 Pa. Calculate the final temperature of the gas. **[2]**

$$\frac{PV}{T} = constant$$

$$\frac{V}{T} = constant\ or\ V \propto T$$ ✔

If the volume increases, so does the temperature by the same factor.

Therefore $T = 483 \times 1.3 = 628\ K$ ✔

Examiner's Commentary

The most common mistake made by candidates is to use the temperature in celsius. In the equation above, T is the thermodynamic temperature in kelvin. There was one mark reserved for the correct conversion of the gas temperature.

Another way to determine the final temperature would be to use
PV = nRT
together with the answer to (a) and the new value for the gas volume. The candidate's answer is brief and elegant. It makes use of the idea that for a constant pressure, Charles' law
V ∝ T
is applicable to the gas.

Exam practice questions

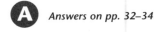

Answers on pp. 32–34

(1) (a) State Boyle's law. [2]

(b) A diver working at a depth of 15 m in the sea releases a rubber balloon to mark his position. At this depth, the volume of the balloon is 4.0×10^{-3} m^3 and the pressure exerted by the gas within the balloon is 2.5×10^5 Pa. The temperature of the sea water is 18°C.

   (i) Show that the amount of gas within the balloon is 0.41 mol.
   Data: R = 8.31 J mol^{-1} K^{-1} [3]

   (ii) Calculate the number of gas molecules within the balloon.
   Data: Avogadro constant, $N_A = 6.02 \times 10^{23}$ mol^{-1} [2]

   (iii) The volume of the balloon increases as it rises towards the surface. At the water surface, the pressure exerted by the gas within the balloon decreases to 1.0×10^5 Pa. Calculate the volume of the balloon at the surface. [2]

(2) (a) Explain in terms of molecular movement the origin of pressure within a container. [4]

(b) For an ideal gas, the mean translational kinetic energy E_k of a molecule is given by
$$E_k = \tfrac{3}{2} kT$$
Define the symbols k and T. [2]

(c) The surface temperature of a star is 5800 K. On its surface, protons move randomly and behave like the molecules of an ideal gas. For these surface protons, calculate

   (i) the mean translational kinetic energy of each proton.
   Data: k = 1.38×10^{-23} J K^{-1} [2]

   (ii) the root-mean-square (r.m.s.) speed of the protons.
   Data: mass of proton, $m_p = 1.7 \times 10^{-27}$ kg. [3]

(d) Explain how the answer to (c) (ii) would change if the particles on the surface of the star were helium nuclei. [2]

(3) (a) Define specific heat capacity of a substance. [1]

(b) A 1.0 kW electric kettle contains 450 g of water at 15°C. The specific heat capacity of the water is 4.2×10^3 J kg^{-1} K^{-1}.

   (i) Calculate the energy supplied by the heating element of the kettle to raise the temperature of the water to 100°C. State any assumption made. [3]

(ii) How long would it take to raise the temperature of the
water to 100°C? **[2]**

(iii) The graph below shows the actual variation of the temperature
θ of the water inside the electric kettle with time t.
(The kettle is switched on at time $t = 0$ s)

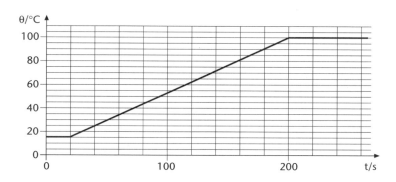

1. Suggest why the temperature of the water remains constant
for the first 20 s. **[1]**

2. What is happening to the energy supplied to the water after 200 s? **[1]**

(4) **(a)** A radioactive material emits alpha (α) particles. What is the nature
of alpha particles? **[2]**

(b) The radioactive source used in many domestic smoke alarms is
americium-241. The source is housed within a plastic case. The
isotopes of americium-241 have a half-life of 460 years. In each decay of
^{241}Am nucleus, an alpha particle of kinetic energy 5.4 MeV is released.

For one particular source in a smoke alarm, the activity is 3.5×10^3 Bq.
Calculate

(i) the decay constant, λ, for the isotopes of ^{241}Am, **[2]**

(ii) the number of ^{241}Am nuclei in the source, **[3]**

(iii) the rate of energy released by the source.
Data: 1 MeV = 1.6×10^{-13} J. **[3]**

(c) Why is it sensible to use an americium-241 source within the
domestic smoke detector rather than a source that emits either beta
particles or γ-rays? **[2]**

Answers

(1) (a) For a fixed amount of ideal gas at constant temperature, the pressure exerted by the gas is inversely proportional to its volume.

(b)(i) $T = 273 + 18 = 291$ K

$PV = nRT$

$2.5 \times 10^5 \times 4.0 \times 10^{-3} =$

$n \times 8.31 \times 291$

$n = 0.414$ mol

> **Examiner's tip**
>
> *It is very important that the temperature is converted into kelvin. Using 18°C would lead to the wrong answer.*

(ii) number $= n \times N_A$

number $= 0.414 \times 6.02 \times 10^{23}$

number $= 2.49 \times 10^{23}$

$\approx 2.5 \times 10^{23}$ molecules

> **Examiner's tip**
>
> *Another route to the correct answer would be to use*
> $$PV = nRT$$
> *together with the value of n from (b)(i). There is not much to choose from between this and the technique shown.*

(iii) $PV = $ constant

(since the temperature is constant)

$2.5 \times 10^5 \times 4.0 \times 10^{-3} =$

$1.0 \times 10^5\ V$

$V = 1.0 \times 10^{-2}$ m^3

(2) (a) Any *four* from:

The molecules are moving about in a random manner.

Each molecule collides with the container wall, resulting in a **change** in velocity (or momentum).

According to Newton's second law, there is a tiny force exerted on the molecule by the wall.

There is an equal but opposite force exerted on the wall by the molecule (Newton's third law).

There are numerous molecular collisions with the walls giving rise to a larger force on the wall.

The pressure P on the wall is given by

$$P = \frac{\text{force on wall}}{\text{area of wall}}.$$

> **Examiner's tip**
>
> *There are four marks for this question. It is important that you do not repeat the same points over and over again. Before answering the question, it is advisable to make a quick list of four distinct points relevant to the question.*

(b) k is Boltzmann constant.

T is the thermodynamic temperature in kelvin.

(c)(i) $E_k = \frac{3}{2} kT$

$E_k = \frac{3}{2} \times 1.38 \times 10^{-23} \times 5800$

$E_k = 1.2 \times 10^{-19}$ J

(ii) $E_k = \frac{1}{2} m \langle c^2 \rangle$

$$\langle c^2 \rangle = \frac{2 \times 1.2 \times 10^{-19}}{1.7 \times 10^{-27}}$$

r.m.s. speed $= \sqrt{1.412 \times 10^8}$

r.m.s. speed $= 1.19 \times 10^4 \approx 1.2 \times 10^4 \text{ ms}^{-1}$

Examiner's tip

*Candidates often confuse the r.m.s. speed ($\sqrt{\langle c^2 \rangle}$) with the mean-square speed ($\langle c^2 \rangle$) of the particles. To avoid this complication, it is sensible to write the mean kinetic energy of the particle as $\frac{1}{2} mv^2$, where **v** is now the **r.m.s. speed** of the particle. The above equation is 'familiar' and does not have any unnecessary or cumbersome notation. The answer would therefore be*

$$\frac{1}{2} mv^2 = 1.2 \times 10^{-19}$$

$$v^2 = \frac{2 \times 1.2 \times 10^{-19}}{1.7 \times 10^{-27}} = 1.412 \times 10^8$$

Hence $v = \sqrt{1.412 \times 10^8} \approx 1.2 \times 10^4 \text{ ms}^{-1}$

(d)   The mean translational kinetic energy is the same.
The more **massive** helium nuclei would move much slower.

Examiner's tip

At the same temperature, the mean kinetic energy E_k of any particle will be the same.

Therefore $\qquad\qquad \frac{3}{2} kT = \frac{1}{2} m \langle c^2 \rangle$

At the same temperature T $\qquad \langle c^2 \rangle \propto \dfrac{1}{m}$.

*Hence, a helium nucleus, which is roughly **four** times more massive than a proton, will move at **half** the r.m.s speed of the proton.*

(3) (a)   Specific heat capacity of a substance is the energy required to change the temperature of a unit mass of the substance by 1K (or 1°C).

(b)(i) The assumption is that there is no loss of heat to the surroundings.
$\Delta E = mc\Delta\theta$
$\Delta E = 0.45 \times 4.2 \times 10^3 \times (100 - 15)$
$\Delta E = 1.61 \times 10^5 \approx 1.6 \times 10^5 \text{ J}$

(ii) $\Delta t = \dfrac{\Delta E}{P}$

$$\Delta t = \frac{1.61 \times 10^5}{1.0 \times 10^3}$$

$\Delta t = 161 \approx 160 \text{ s}$

(iii) 1. The heating element of the kettle is being heated instead of the water.

2. The water is boiling. The energy supplied to the water is used to break molecular bonds as the water changes state into steam.

Answers

(4) (a) Any *two* from:
Alpha particles are:
helium **nuclei**,
positively charged
and consist of two protons and two neutrons.

(b)(i) 1 year = $365 \times 24 \times 3600 = 3.154 \times 10^7$ s

$$\lambda = \frac{\ln(2)}{T_{\frac{1}{2}}} \approx \frac{0.693}{T_{\frac{1}{2}}}$$

$$\lambda = \frac{0.693}{460 \times 3.154 \times 10^7}$$

$$\lambda = 4.78 \times 10^{-11} \approx 4.8 \times 10^{-11} \text{ s}^{-1}$$

> **Examiner's tip**
>
> *It is important that the answer is given in S.I. units. It would therefore be sensible to convert the half-life into seconds.*

(ii) The decay equation is
$$\frac{\Delta N}{\Delta t} = -\lambda N \text{ or } A = -\lambda N$$
where A is the rate of decay of nuclei (or the activity of the source).

$$N = \frac{A}{\lambda} \qquad \text{(Ignore the minus sign)}$$

$$N = \frac{3.5 \times 10^3}{4.78 \times 10^{-11}}$$

$$N = 7.32 \times 10^{13} \approx 7.3 \times 10^{13} \text{ nuclei}$$

(iii) K.E. of each α particle = 5.4 MeV
K.E. of each α particle = $5.4 \times 1.6 \times 10^{-13} = 8.64 \times 10^{-13}$ J

Power = rate of energy release
Power = activity × energy of each α particle
$P = 3.5 \times 10^3 \times 8.64 \times 10^{-13}$
$P = 3.02 \times 10^{-9} \approx 3.0 \times 10^{-9}$ Js^{-1} (W)

> **Examiner's tip**
>
> *This question requires a good understanding of the term **activity** which represents the number of emissions from the source per unit time. Since power is the amount of energy released per unit time, it follows that: power = activity × energy of each α particle.*

(c) It is safer.
This is because the α particles cannot penetrate the plastic case of the alarm but β particles and γ-rays can.

Questions with model answers

C grade candidate – mark scored 6/9

For help see Revise AS Study Guide pages 130 and 131

(1) Explain what is meant by a photon.   **[1]**

> *A photon is how energy is carried by an electromagnetic wave.*
>
> *It is a quantum (or packet) of energy.*   ✔
>
> *A photon travels at the speed of light.*

Examiner's Commentary

This is a good answer by the candidate. There is only one mark available and the candidate has done more than was required. The crucial response that a photon is a 'quantum of energy' was awarded the one mark by the examiner. The last statement is correct, but there are no extra marks that could be given for this interesting comment.

(2) A light-emitting diode (LED) emits visible light when it conducts. For one particular LED, it **just** starts to emit red light when the p.d. across it is 1.8 V and the current through it is 1.2 mA. The light emitted by the LED has a wavelength of 6.8×10^{-7} m.

(a) What is the input electrical power to the LED?   **[2]**

> $P = VI$
>
> $P = 1.8 \times 1.2$ ✔
>
> $P = 2.16 \ mW$ ✔

The candidate secured the second mark because the correct prefix of the 'milli' was inserted in front of the unit for power. It is always safer to substitute and write answers in standard form.

(b) Calculate the energy of each photon of red light from the LED.
Data: $h = 6.63 \times 10^{-34}$ Js
$\quad\quad c = 3.0 \times 10^8$ ms^{-1} **[3]**

> $E = hf$ ✔
>
> $E = \dfrac{h\lambda}{c}$ ✗
>
> $E = \dfrac{6.63 \times 10^{-34} \times 6.8 \times 10^{-7}}{3.0 \times 10^8} = 1.5 \times 10^{-48} \ J$ ✗

The first mark was awarded for correctly recalling the equation for the energy of a photon. The candidate's second step is wrong. Since
$c = f\lambda$ *and* $E = hf$
then $E = \dfrac{hc}{\lambda}$.
Since the incorrect equation was used by the candidate, no further marks could be awarded. The correct answer is 2.9×10^{-19} J.

(c) Use your answers to **(a)** and **(b)** to estimate the rate at which photons are emitted from the LED. State any assumption made.   **[3]**

> *I am going to assume that all the electrical power goes into producing light.*   ✔
>
> $P = \dfrac{\Delta E}{\Delta t}$
>
> When $t = 1 \ s$ $\Delta E = 2.16 \ mJ$
>
> *Number of photons* $= \dfrac{2.16}{1.5 \times 10^{-48}}$ ✗
>
> *Number of photons* $= 1.4 \times 10^{48}$ *per second.*   ✔

*The candidate has used the wrong answer from **(b)**. This, by itself would have been alright here. Examiners try not to penalise twice for the same error. However, another mistake in the form of the missing 10^{-3} factor for the power has slipped in. The examiner has deducted one mark for this error, but subsequent marks have been awarded. The correct answer for the rate of photon emission from the LED would have been 7.4×10^{15} s^{-1}.*

Questions with model answers

A grade candidate – mark scored 5/5

 For help see Revise AS Study Guide pages 117 and 120

Examiner's Commentary

(1) Explain what is meant by a **transverse** wave. **[1]**

> *A transverse wave has oscillations that are at right angles to the wave velocity.* ✔

(2) According to a student

'Light reflected from the surface of water is **plane polarised**.'

(a) State what is meant by **plane polarised**. **[1]**

> *This is when a wave (transverse) has oscillations in only one plane.* ✔

(b) Name a wave, other than visible light, that can be plane polarised. **[1]**

> *All electromagnetic waves can be polarised.* ✔
>
> *All the following waves can be polarised:*
>
> *γ-rays, X-rays, ultra-violet light, infra-red light, microwaves and radio waves.*

This is a superb answer. The response is typical from a grade A candidate. The candidate has secured the mark by stating that all electromagnetic waves can be polarised, but has elaborated a bit more, just to be on the safe side.

(c) Outline an experiment to assess the validity of the student's statement. **[2]**

> *I would use two polarising filters and look through both filters with my eye.* ✔
>
> *One filter is fixed and the other is rotated in front of the other.*
>
> *If the light is polarised, then the intensity of light would change from maximum to zero.* ✔

Since the reflected light from the water is thought to be plane polarised, it is possible to use just one polarising filter rotated in front of the eye. If the student's statement is true, then the experimental observations would be the same as that outlined by the candidate.

Exam practice questions

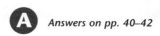 *Answers on pp. 40–42*

(1) **(a)** State three differences between a sound wave and a light wave. **[3]**

(b) Sound waves are produced in air by a loudspeaker connected to a signal generator. The frequency of sound is increased.
State the effect, if any, of this increase on

    (i) the speed of sound, **[1]**

    (ii) the wavelength of the wave. **[1]**

(c) Calculate the frequency required in **(b)** for the waves to have a wavelength of 84 cm when the speed of sound in air is 340 ms^{-1}. **[2]**

[OCR Specimen, 2000]

(2) **(a)** The diagram shows a ray of light within a semi-circular block of glass making an angle of incidence equal to the **critical angle**, c for glass.

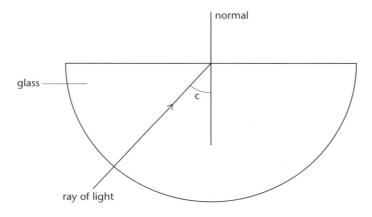

    (i) Complete the diagram above to show what happens to the ray of light. **[2]**

    (ii) The critical angle, c for the glass is 42°. Calculate the refractive index for the glass. (You may assume that air has a refractive index of 1.00.) **[2]**

(b) The diagram below shows the cross-section through an optical cable made of glass of refractive index 1.52. A ray of light incident at an angle of 20° to the glass-air interface is internally reflected. The total length of the optical cable is 1.20 km.

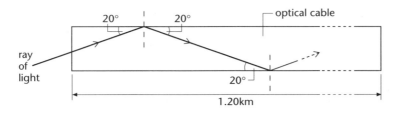

    (i) Calculate the speed of light in the optical cable.
    Data: $c = 3.00 \times 10^8$ ms^{-1} **[2]**

Exam practice questions

(ii) For a ray of light incident at an angle of 20° to the glass-air interface, calculate the time taken for it travel the total length of the cable. **[3]**

(iii) State how the answer to **(b)(ii)** would change for a ray of light travelling along the axis of the cable. **[1]**

(iv) Suggest why it may be sensible for an optical cable to have a very small diameter when transmitting digital information. **[1]**

(3) **(a)** Outline two properties of a progressive wave. **[2]**

(b) The diagram shows two identical loudspeakers emitting sound when connected to a signal generator set at 1.2 kHz.

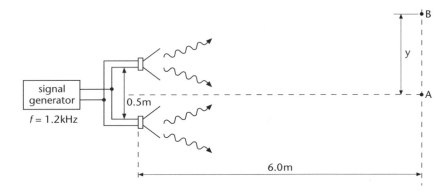

(i) Calculate the wavelength of sound emitted from each loudspeaker.
Data: Speed of sound in air = 340 ms^{-1} **[2]**

(ii) A student is listening to the sound from **both** loudspeakers.
When at A, a very loud sound is heard. As the student slowly moves towards B, the intensity of sound gradually decreases. When at B, virtually no sound is heard.
Explain why a loud sound is heard at point A and virtually none at the adjacent point B. **[2]**

(iii) Calculate the distance *y* between A and B. **[3]**

(iv) State and explain how the answer to **(b)(iii)** would change if the frequency of sound were to be doubled. (You are not expected to do any further calculations.) **[2]**

(4) **(a)** Outline some of the main properties of electromagnetic **waves**. Name one principal region of the electromagnetic spectrum and suggest a practical application for the waves. **[5]**

(b) A gold-leaf electroscope consists of a metal cap and a thin strip of gold foil attached at the end of the metal stem. The diagram shows a negatively charged electroscope, with the gold-leaf diverged.

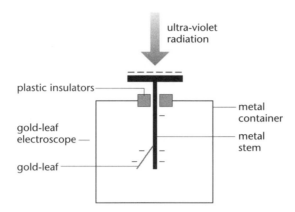

The cap of the electroscope is made of the metal zinc. When the cap is exposed to a weak ultra-violet source, the gold-leaf starts to collapse. After some time, it shows no divergence.

Outline the phenomenon of the photoelectric effect and use the ideas developed to explain why the divergence of the gold-leaf decreases with time.

[5]

Answers

(1) (a)   Sound requires a medium and light does not.
Speed of sound is much slower.
Sound is a longitudinal wave and light is a transverse wave.

(b)(i)   No effect on the speed of sound.

   (ii)   The wavelength of sound decreases.

Examiner's tip

The speed of sound, v, is a constant. The speed is given by the equation

$$v = f\lambda$$

As the frequency is increased, the wavelength must decrease because the product fλ is a constant. The same conclusion can be arrived at through

$$\lambda \propto \frac{1}{f}.$$

(c) $v = f\lambda$

$$f = \frac{340}{0.84}$$

$$f = 405 \text{ Hz.}$$

Examiner's tip

Remember to convert the wavelength into metres.

(2) (a)(i)   Ray is refracted at 90°.
There is a weak reflection within the glass block.

   (ii) $n = \dfrac{1}{\sin c}$
where n = refractive index and c is the critical angle within the glass.

$$n = \frac{1}{\sin 42°}$$

$$n = 1.494 \approx 1.49$$

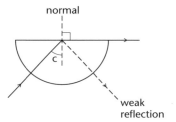

Examiner's tip

According to Snell's law $\dfrac{\sin i}{\sin r} = {}_1n_2$

*As the ray of light travels towards the glass-air interface, an incidence angle of 42° within **glass** gives a refracted angle of 90° in **air**. Therefore*

$$\frac{\sin 42°}{\sin 90°} = {}_{glass}n_{air}$$

But ${}_{air}n_{glass} = \dfrac{1}{{}_{glass}n_{air}}$

$$\therefore {}_{air}n_{glass} = \frac{1}{\sin 42°} \approx 1.49$$

(b)(i) $n = \dfrac{c_v}{c_m}$

$c_m = \dfrac{3.00 \times 10^8}{1.52}$

$c_m = 1.974 \times 10^8 \approx 1.97 \times 10^8 \ ms^{-1}$

(ii) distance $= \dfrac{1.2 \times 10^3}{\cos 20°} = 1.277 \times 10^3 \ m$

time $= \dfrac{\text{distance}}{\text{velocity}}$

$t = \dfrac{1.277 \times 10^3}{1.974 \times 10^8}$

$t \approx 6.47 \times 10^{-6} \ s$

Examiner's tip

The actual 'zigzag' path of the reflected ray can be stretched out as shown below.

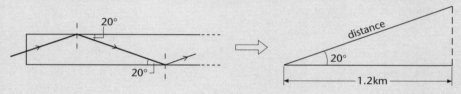

The total distance travelled by the internally reflected ray is given by

$$\cos 20° = \dfrac{1.2 \times 10^3}{\text{distance}} \ .$$

In order to determine the time taken, you must use the actual velocity of the ray within the optical cable. To use a velocity of $3.00 \times 10^8 \ ms^{-1}$ would be incorrect.

(iii) The time taken would be shorter.

Examiner's tip

The reason for the shorter time is because the distance travelled by the ray would be less than the longer 'zigzag' path.

(iv) There would be reduced signal 'smearing'.

(3) (a) Any *two* from:
A progressive wave travels through space carrying energy.
It carries energy through space via vibrations (or oscillations).
It has wavelength, frequency and wave velocity.
The wave velocity, v, is given by $v = f\lambda$.

Answers

(b)(i) $v = f\lambda$

$$\lambda = \frac{340}{1.2 \times 10^3}$$

$\lambda = 0.283 \approx 0.28$ m

(ii) At A, the waves from the loudspeakers interfere **constructively**.
At B, the waves from the loudspeakers interfere **destructively**.

(iii) $x = \dfrac{\lambda D}{a}$

$$x = \frac{0.283 \times 6.0}{0.5}$$

$y = \dfrac{x}{2}$

$y = 1.70 \approx 1.7$ m.

(iv) The wavelength of sound would be halved.
Therefore, y will be reduced by a factor of two.

Examiner's tip

The separation x between neighbouring maximum (or minimum) signals is given by the expression $x = \dfrac{\lambda D}{a}$.
In the question, the required distance is between the maximum signal at A and the adjacent minimum signal at B. Therefore, the distance y will be half that of x.

Examiner's tip

There are two main ideas here. They are $\quad v = f\lambda$ *and* $x = \dfrac{\lambda D}{a}$.

$$\therefore \; x = \frac{Dv}{af}$$

$$\text{or } x \propto \frac{1}{f}.$$

*The separation x, and hence y, will be **halved** when the frequency of the signal is **doubled**.*

(4) (a) Electromagnetic waves are:
Transverse waves that travel through a vacuum at a velocity of 3.0×10^8 ms^{-1}.
E.M. waves consist of oscillating electric and magnetic fields.

One of the principal regions of the electromagnetic spectrum is the infra-red region.
Infra-red radiation is used by television remote controls and security lighting at night.

Examiner's tip

*The principal regions of the electromagnetic spectrum in order of **increasing** wavelength are: γ-rays, X-rays, ultra-violet, visible light, infra-red, microwaves and radio waves. The practical applications are too numerous to list. Microwaves, for example, are used in microwave ovens and for mobile phone communication.*

(b) Negatively charged electrons are removed from the surface of zinc by the **photons** interacting with these electrons.
As a result the divergence of the gold-leaf decreases.
Each photon interacts with a single surface electron.
Energy is conserved, therefore '$hf = \phi + K.E_{max}$'.
Electrons are removed because the energy of ultra-violet photons is greater than the work function ϕ of the metal.

Questions with model answers

C grade candidate – mark scored 6/10

For help see Revise A2 Study Guide chapters 1.1 and 1.2

(1) A jet of water from a hose pipe hits a wall at right angles to it and then trickles down the wall.

(a) Explain how the water jet exerts a force on the wall. **[3]**

> *The water jet has mass and velocity and therefore momentum.* ✔
>
> *It hits the wall with a force.* ✗ ←
>
> *The jet loses its power and then trickles down the wall.* ✗

(b) Calculate the magnitude of the force exerted on the wall by the water jet if the water is delivered at a rate of 8.0 kgs⁻¹ and hits the wall with a velocity of 25 ms⁻¹. **[2]**

> $$F = \frac{m(v-u)}{t}$$ ✔
>
> $$F = \left(\frac{m}{t}\right) \times u \quad \text{(magnitude only)}$$
>
> $$F = 8.0 \times 25 = 200 \text{ N}$$ ✔

(2) A tennis player hits an incoming ball of mass 60 g with a racquet and changes its direction of travel. The force exerted by the racquet on the ball is 150 N. The diagram shows how the velocity of the ball is changed by the racquet.

     30ms⁻¹   50ms⁻¹

      Before   After

(a) Calculate the magnitude of the impulse exerted by the racquet **[2]**

> *Impulse* $= F\Delta t$
>
> *Impulse* $=$ *change in momentum* ✔
>
> *Impulse* $= (0.06 \times 50) - (0.06 \times 30)$ ✗
>
> *Impulse* $= 1.2$ Ns

(b) Hence calculate the time of impact of the ball with the racquet **[3]**

> *Impulse* $=$ *change in momentum of ball*
>
> $F\Delta t = 1.2$ ✔   ⟨error carried forward⟩
>
> $\Delta t = \dfrac{1.2}{150}$ ✔
>
> $\Delta t = 8.0 \times 10^{-3}$ ✗

Examiner's Commentary

*This is a disappointing answer. One mark has been awarded for the first statement, but the other two statements are either incomplete or misleading. On impact with the wall, there is a **change** in momentum for the water jet. According to Newton's second law, a change in momentum in a given time results in a force on the water jet. This force is provided by the wall. The water jet exerts an equal but opposite force on the wall (Newton's third law). The magnitude of the force depends on the velocity of the water jet and on the rate at which the mass of water hits the wall.*

The velocity of the water after impact is zero. Hence, the candidate has correctly made the final velocity 'v' equal to zero.

Momentum is a vector quantity and as such, it has both magnitude and direction. The momentum of the ball after impact with the racquet has to be of the opposite sign to the initial momentum. The magnitude of the impulse provided by the racquet is

impulse $= (0.06 \times 50) - (0.06 \times -30)$

impulse $= 4.8$ kgms⁻¹

The candidate has used the wrong answer of 1.2 Ns from (a). This mistake has already been penalised. The examiner has been fair and awarded all subsequent marks. Sadly, the candidate lost the final mark because the unit for time has been missed out.

Questions with model answers

A grade candidate – mark scored 8/8

For help see Revise A2 Study Guide chapters 1.1, 1.2 and 1.3

Examiner's Commentary

(1)   Explain how the tyres of a car generate a forward force on the car. **[3]**

> *The car tyres rotate* ✔
> *in such as way that they exert a **backward** force on the ground.* ✔
> *According to Newton's third law, `action = reaction', therefore*
> *the ground exerts an equal **forward** force on the tyres.* ✔

(2)   A car and its occupants of total mass of 1.2×10^3 kg is moving along the motorway. At a particular time, its speed and acceleration are 15 ms^{-1} and 0.2 ms^{-2} respectively. At this time the total resistive force on the car is 150 N.

(a)   Name one of the resistive forces on the car.   **[1]**

> *As the car moves through the air, the air exerts a*
> *resistive force called drag.* ✔

(b)   Calculate the resultant force on the car.   **[2]**

> $F = ma$ ✔
> $F = 1.2 \times 10^3 \times 0.2$
> $F = 240\ N$ ✔

(c)   Calculate the motive power developed by the car.   **[2]**

> $P = Fv$
> $F = 150 + 240 = 390\ N$ ✔
> $Power = 390 \times 15 = 5.9\ kW$ ✔

*The motive force provided by the car is 390 N. In order to determine the power developed by the car, it would be incorrect to use just the resultant force from (**b**).*

Exam practice questions

Ⓐ *Answers on pp. 47–48*

(1) **(a)** The diagram shows a rocket travelling in space.

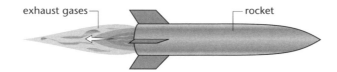

exhaust gases — rocket

Use your knowledge of Newton's laws to explain the origin of the force on the rocket as it expels exhaust gases at high velocity. **[4]**

(b) A bullet of mass 1.40×10^{-2} kg is fired horizontally from a gun with a velocity of 210 ms⁻¹. It hits and gets embedded inside a stationary wooden block. The block of wood has a mass of 1.50 kg and lies on a horizontal frictionless surface. After impact, the wooden block (together with the embedded bullet) moves with a constant velocity.

(i) Calculate the momentum of the bullet just before it enters the wooden block. **[2]**

(ii) Calculate the velocity, v of the wooden block after being hit by the bullet. **[3]**

(iii) Explain whether or not the kinetic energy of the bullet is conserved. **[2]**

(2) The diagram shows a 70 kg athlete running up a hill that is sloping at an angle of 6.0° to the horizontal. The athlete maintains a constant speed whilst running up the hill.

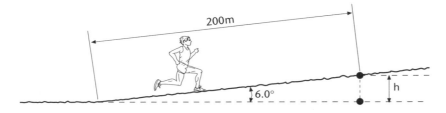

200m

6.0°

h

In 50 s, the athlete covers 200 m along the hill. Calculate

(a) the vertical distance h climbed by the athlete, **[2]**

(b) the gravitational potential energy gained by the athlete,
Data: g = 9.8 Nkg⁻¹ **[2]**

(c) the power developed by the athlete. **[2]**

Exam practice questions

(3) **(a)** Define work done by a force. [2]

(b) An object is **released** close to the surface of the Earth.

 (i) Use the idea of work done by a force, to explain why the speed of the object increases as it falls towards the ground. [3]

 (ii) Explain whether or not momentum is conserved as the object accelerates towards the ground. [2]

(c) A swing consists of a rubber tyre of mass 14 kg suspended from a 6.0 m long rope. A child of mass 35 kg sits on the swing. The speed of the child at the lowest point is 4.0 ms^{-1}.

 (i) Determine the centripetal acceleration of the child at this lowest point. [2]

 (ii) Use your answer to **(c)(i)** to determine the tension T in the rope at this lowest point.
 Data: g = 9.8 Nkg^{-1} [3]

Answers

(1) (a) The exhaust gases have momentum.
A rate of change of momentum takes place for the exhaust gases.
According to Newton's second law, there is a force exerted on the exhaust gases.
According to Newton's third law, the force acting **on** the exhaust gases is equal and opposite to that experienced **by** the rocket.
Hence there is a force acting on the rocket.

(b)(i) $p = mv$
$p = 1.40 \times 10^{-2} \times 210$
$p = 2.94$ kgms^{-1}

(ii) Total mass after impact $= M$
$M = 1.50 + 0.014 = 1.514$ kg
Momentum is conserved in the 'collision'.
$\therefore\ 2.94 = 1.514\ v$
$v = 1.94$ ms^{-1}

(iii) Kinetic energy of the bullet is **not** conserved.
Some of the kinetic energy of the bullet is converted into heat, sound, etc.
The impact is an inelastic collision.

> **Examiner's tip**
>
> *In all collisions, where there are no external forces, **momentum** is always conserved. Hence*
> *Initial momentum = final momentum*
> *In this collision, **energy** is also conserved. However, kinetic energy is not. Some of the kinetic energy of the bullet is used to make a dent in the wood, generate sound etc. So, it would be wrong to suggest that kinetic energy is conserved in this impact. You <u>cannot</u> therefore write the following:*
> $$\text{initial K.E.} = \text{final K.E.}$$
> $$\tfrac{1}{2} \times 1.40 \times 10^{-2} \times 210^2 = \tfrac{1}{2} \times 1.514 \times v^2.$$
> *The expression above may only be written for an **elastic collision**, where kinetic energy is conserved.*

(2) (a) $\sin 6.0° = \dfrac{h}{200}$
$h = 200 \times \sin 6.0°$
$\therefore\ h = 20.9 \approx 21$ m

(b) Gain in gravitational potential energy = G.P.E.
G.P.E. $= mg\Delta h$
G.P.E. $= 70 \times 9.8 \times 20.9$
G.P.E. $= 1.43 \times 10^4 \approx 1.4 \times 10^4$ J

(c) Power $= \dfrac{\text{energy transfer}}{\text{time taken}}$

$P = \dfrac{1.43 \times 10^4}{50}$

$P = 286 \approx 290$ W

(3) (a) Work done is defined as the product of the force and the distance moved by the force.
The distance moved is that in the **direction** of the force.

(b)(i) The weight of the object accelerates it towards the Earth.
Work is done on the object by the weight.
Work done by the weight = gain in kinetic energy.
Therefore the speed of the falling object increases during its descent.

Answers

(ii) The falling object exerts an equal but opposite force on the Earth.
The Earth acquires an **equal** but **opposite** momentum to the falling object.
The total momentum is zero.
Momentum is therefore conserved.

Examiner's tip

It is very important to appreciate the idea that momentum is a vector quantity.

(c)(i) $a = \dfrac{v^2}{r}$

$a = \dfrac{4.0^2}{6.0}$

$a = 2.67 \approx 2.7 \text{ ms}^{-2}$

(ii) m = total mass
$m = 35 + 14 = 49$ kg
$T - mg = ma$
$T = ma + mg$
$T = 49 \times (2.67 + 9.8)$
$T = 611 \approx 610$ N

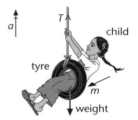

Examiner's tip

It is very tempting to calculate the tension T in the rope by using
$$T = ma.$$
*Such an approach would be wrong. The **resultant** force at the lowest point of the swing is*
$$T - weight.$$
Once the resultant force has been identified, Newton's second law may be used to calculate the tension T in the rope. Therefore
$$T - mg = ma$$

Questions with model answers

C grade candidate – mark scored 4/6

 For help see Revise A2 Study Guide chapter 2.3

(1) State **one** common feature of all electromagnetic waves travelling through a vacuum. **[1]**

> *All E.M. waves travel at the same speed in a vacuum.* ✔

(2) Two radio transmitters broadcast signals at frequencies of 198 kHz and 102 MHz.

(a) State which of the two signals has a longer wavelength. **[1]**

> *The 198 kHz signal.* ✔

(b) Calculate the ratio: **[2]**

$$\frac{\text{wavelength of 198 kHz signal}}{\text{wavelength of 102 MHz signal}}$$

> *The calculation cannot be done because the speed is not given.* ✗

(c) The diagram shows a small community in a hilly region receiving signals from the two transmitters.

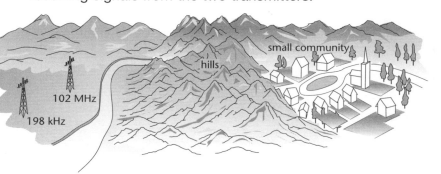

Explain why there are no problems receiving the signals at the longer wavelength. **[2]**

> *The longer wavelength signals are easily diffracted by the hills.* ✔
>
> *Hence the signal can travel round `corners´. It is easily diffracted.* ✔

Questions with model answers

A grade candidate – marks scored 9/10

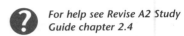

For help see Revise A2 Study Guide chapter 2.4

(1) Outline two ways in which a standing (or stationary) wave differs from a progressive wave.

> *Energy of the wave is localised in space.* ✔
>
> *Points on either side of the nodes oscillate with a phase difference of 180°.* ✔ **[2]**

(2) A loudspeaker, connected to a signal generator set at 2.0 kHz, is placed in front of a smooth flat vertical wall. This is illustrated in the diagram below.

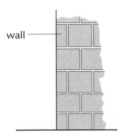

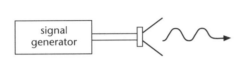

wall

signal generator

(a) Explain how a standing wave is set up between the loudspeaker and the wall. **[2]**

> *Waves from the loudspeaker combine or interfere* ✔
> *with the identical (coherent) waves reflected from the wall.* ✔
> *The result of this is a standing wave pattern between the wall and the loudspeaker.*

(b) A small microphone moved from the loudspeaker towards the wall detects a series of maxima and minima. Calculate

(i) the wavelength λ of the sound emitted by the loudspeaker
Data: speed of sound = 340 ms⁻¹ **[2]**

$$\lambda = \frac{340}{2.0 \times 10^3}$$ ✔
$$\lambda = 0.17 \text{ m}$$ ✔

(ii) the distance between successive maxima. **[2]**

> *separation between adjacent nodes (or antinodes)* $= \dfrac{\lambda}{2}$ ✔
>
> *distance between maxima* $= \dfrac{0.17}{2}$
>
> *distance* $= 8.5 \text{ cm}$ ✔

(3) State and explain how your answer to **2(b)(ii)** would change if the frequency of sound is increased. **[2]**

> *The distance between the maxima will decrease.* ✔ ←

The answer is brief, with no explanation given for the statement. The wavelength λ decreases as the frequency f increases because

$$\lambda \propto \frac{1}{f}$$

Since the separation between adjacent maxima is $\dfrac{\lambda}{2}$, this distance will therefore also decrease.

Exam practice questions

 Answers on pp. 54–56

(1) **(a)** Explain what is meant by resonance. Draw a sketch graph of amplitude
A of a mechanical oscillator against the forcing frequency *f* to illustrate
your answer. **[4]**

(b) The molecules of HF absorb infra-red radiation very strongly at a
wavelength of 2.4×10^{-6} m. Use this information to calculate the
natural frequency f_0 with which the molecules of HF vibrate.
Data: $c = 3.0 \times 10^8 \ \text{ms}^{-1}$. **[3]**

(2) **(a)** Define the **period** of a mechanical oscillator. **[1]**

(b) Define simple harmonic motion (s.h.m.). **[2]**

(c) The graph below shows the variation of the displacement *x* with time *t*
for an object executing s.h.m. The mass of the object is 3.0 kg.

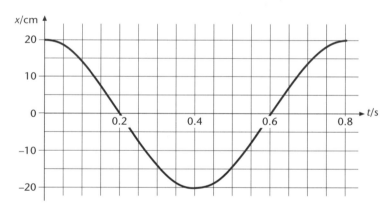

(i) What is the amplitude of the motion? **[1]**

(ii) On the graph, indicate with a cross (**✗**), the time at which the object
has maximum speed. **[1]**

(iii) Calculate the maximum speed of the object. **[3]**

(iv) Calculate the maximum force acting on the object. **[3]**

(v) State one way in which the shape of the graph would change if
there was a small amount of damping. **[1]**

Exam practice questions

(3) The diagram shows a long horizontal plastic tube containing some fine powder. One end of the tube is closed and a loudspeaker is positioned at the other end. The loudspeaker is connected to a signal generator.

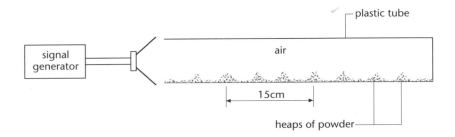

At a particular frequency, a standing wave is set up within the tube. The powder within the tube forms heaps at the **nodes**.

(a) State why the powder forms heaps at the **nodes**. [1]

(b) Determine the wavelength λ of the standing wave. [3]

(c) Calculate the frequency of the sound produced by the loudspeaker.
Data: Speed of sound = 340 ms^{-1} [2]

(d) Suggest why the powder disperses when the frequency of sound is altered slightly. [1]

(4) **(a)** What is a diffraction grating? [1]

(b) Write an equation for the condition for diffraction maximum for a diffraction grating. Define any symbols used. [2]

(c) Blue and red light of wavelengths 4.6×10^{-7} m and 6.7×10^{-7} m respectively is incident at right angles on a diffraction grating that has 600 lines per mm.

   (i) Show that the grating spacing is about 1.7×10^{-6} m. [1]

   (ii) Calculate the angle between the second-order maxima for the red and the blue light. [3]

   (iii) State and explain how your answer to **(c)(ii)** would change if the light was incident at an angle as shown in the diagram below.

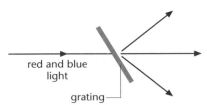

[2]

(5) Electromagnetic radiation is incident on a metal surface. The results from an experiment are shown on the graph below. The maximum kinetic energy of the photoelectrons is E_k and the frequency of the incident electromagnetic radiation is f.

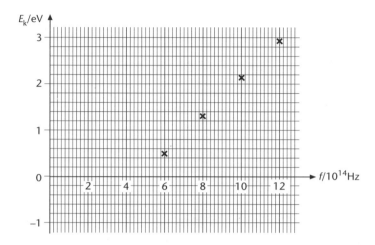

(a) Draw a line of best-fit. **[1]**

(b) State why there are no data points for negative values for E_k. **[1]**

(c) Use the graph to determine the threshold frequency f_0. **[1]**

(d) Calculate the work function ϕ of the metal in joules.
Data: $h = 6.63 \times 10^{-34}$ Js **[3]**

(e) Explain why the gradient of the line is equal to Planck's constant h. **[2]**

Answers

(1) (a) At resonance, the oscillator has maximum amplitude
and the forcing frequency is equal to the natural frequency of the oscillator.
The oscillator absorbs maximum energy at resonance.

Correct shape of the A against f sketch.

(b) At resonance, the frequency of the infra-red radiation is equal to the natural
frequency of HF molecules.

$$f_0 = \frac{c}{\lambda}$$

$$f_0 = \frac{3.0 \times 10^8}{2.4 \times 10^{-6}}$$

$$f_0 = 1.25 \times 10^{14} \approx 1.3 \times 10^{14} \text{ Hz}$$

Examiner's tip

*The oscillating electric field of the infra-red radiation is responsible for **forcing** the HF molecules to vibrate. At resonance, the HF molecules absorb the maximum amount of energy from the incident infra-red radiation.*

(2) (a) The period is the time taken for one complete oscillation.

(b) The acceleration of the object $\propto$ displacement from some fixed point.
It is directed towards some fixed point.
Therefore, $a \propto -x$.

(c)(i) Amplitude = 20 cm.

Examiner's tip

The most common mistake made by candidates is to give the peak-to-peak distance. It is important to be aware that amplitude is equal to the maximum displacement from the equilibrium position of the oscillator.

(ii) A cross (✗) at either $t = 0.2$ s or $t = 0.6$ s.

Examiner's tip

The gradient from a displacement-time graph is equal to the velocity of the oscillator. The gradient is maximum at the times indicated above. At 0.2 s the object is moving one way and after one half of an oscillation, it is travelling in the opposite direction with the same speed.

(iii) $T = 0.8$ s

$$\omega = \frac{2\pi}{T} = \frac{2\pi}{0.8}$$

$$\omega = 7.854 \text{ rad s}^{-1}$$

$$v_{max} = \omega A = 7.854 \times 0.2$$

$$v_{max} = 1.57 \approx 1.6 \text{ ms}^{-1}$$

(iv) $a_{max} = \omega^2 A$

$a_{max} = 7.854^2 \times 0.2$

$a_{max} = 12.34 \text{ ms}^{-2}$

$F = ma = 3.0 \times 12.34$

$F = 37.0 \approx 37 \text{ N}$

(v) The amplitude of the object will decrease.

For a small amount of friction, the amplitude will decay exponentially with respect to time.

(3) (a) There is no oscillation of air particles at the nodes.

Examiner's tip

The fine powder will move around and gather at the points where there is locally no oscillations of the air. It is at these points that the powder gathers and forms tiny heaps.

(b) Separation between successive nodes (or anti-nodes) $= \dfrac{\lambda}{2}$

$\dfrac{3\lambda}{2} = 15 \text{ cm}$

$\therefore \lambda = 10 \text{ cm}$

(c) $v = f\lambda$

$f = \dfrac{340}{0.1}$

$f = 3.4 \times 10^3 \text{ Hz} \ (3.4 \text{ kHz})$

Examiner's tip

For a standing wave, its frequency is equal to the frequency of the sound from the loudspeaker.

(d) There is no standing wave produced within the plastic tube.

(4) (a) A piece of glass or metal with a large number of parallel 'lines' or slits.

(b) $d \sin \theta = n\lambda$

d = grating spacing, θ = angle at which diffraction maximum is observed, n = order of the image and λ = wavelength of incident light.

(c)(i) $d = \dfrac{1\text{mm}}{600}$

$d = \dfrac{1.0 \times 10^{-3}}{600}$

$d = 1.67 \times 10^{-6} \text{ m}$

Examiner's tip

Remember you cannot score a mark without showing how the answer is arrived at. Do not simply put down the last statement.

Answers

(ii) $d \sin \theta = 2 \lambda$ (n = 2)

For red light $\Rightarrow \sin \theta = \dfrac{2 \times 4.6 \times 10^{-7}}{1.67 \times 10^{-6}}$

$\therefore \theta = \sin^{-1}(0.5509) = 33.4°$

For blue light $\Rightarrow \sin \theta = \dfrac{2 \times 6.7 \times 10^{-7}}{1.67 \times 10^{-6}}$

$\therefore \theta = \sin^{-1}(0.8024) = 53.4°$

Angle between the red and blue light $= \Delta\theta$
$\Delta\theta = 53.4 - 33.4 = 20°$

(iii) $\Delta\theta$ will be larger
because the effective value for d is smaller.

Examiner's tip

The separation between neighbouring stretched out fingers of the hand when looking at them at 90° is more than when viewed at an angle. The same happens to the separation between neighbouring slits of the diffraction grating.

(5) (a)   A line of best fit is drawn.

(b)   Photoelectrons are **not** released from the metal surface by the incident radiation.

(c) $f_0 \approx 4.8 \times 10^{14}$ Hz

Examiner's tip

At the threshold frequency, the photoelectrons are just released from the metal surface by the incoming photons. The freed electrons have no kinetic energy. The intercept of the line with the frequency axis is therefore the value of the threshold frequency.

(d) $hf_0 = \phi$
$\phi \approx 6.63 \times 10^{-34} \times 4.8 \times 10^{14}$
$\phi \approx 3.2 \times 10^{-19}$ J

(e) $hf = \phi + E_k$ (Einstein's photoelectric equation)
$\therefore E_k = hf - \phi$
Comparing this with $y = mx + c$, the equation for a straight line,
the gradient must be h, Planck's constant.

Questions with model answers

C grade candidate – mark scored 6/10

 For help see Revise A2 Study Guide chapters 1.4, 3.1 and 3.5

(1) Define gravitational field strength, *g* at a point in space. **[1]**

g is the force experienced <u>per</u> unit mass at a point in a gravitational field. ✔

(2) The mass of the Earth is 6.0×10^{24} kg and its equatorial radius is 6.4×10^6 m.

(a) Calculate the surface gravitational field at the equator.
Data: $G = 6.67 \times 10^{-11}$ Nm^2kg^{-2} **[3]**

$$g = \frac{-GM}{r^2}$$ ✔

$$g = \frac{6.67 \times 10^{-11} \times 6.0 \times 10^{24}}{(6.4 \times 10^6)^2}$$ ✔

$$g = 9.77 \ ms^{-2} \ (or \ Nkg^{-1})$$ ✔

(b) A person at the equator has an orbital period of 1 day (8.64×10^4 s). Calculate, for this person **[2]**

(i) the orbital speed,

$$v = \frac{distance}{time}$$

$$v = \frac{6.4 \times 10^6}{8.64 \times 10^4}$$ ✘

$$v = 74 \ ms^{-1}$$ ✘

(ii) the centripetal acceleration. **[2]**

$$a = \omega^2 r$$

$$a = \left(\frac{2\pi}{8.64 \times 10^4}\right)^2 \times 6.4 \times 10^6$$ ✔

$$a = 3.38 \times 10^{-2} \ ms^{-1}$$ ✔

(c) Calculate the force exerted by the ground on a 70 kg person at the equator. **[2]**

$$force = weight = 700 \ N$$ ✘

Examiner's Commentary

The orbital speed for an object moving in a circle is not equal to the radius of the orbit divided by the period. The speed v is given by

$$v = \frac{circumference \ of \ circle}{time}$$

$$v = \frac{2\pi r}{T} \ or \ v = \omega r$$

Correct substitution into the above equation gives an orbital speed of about 470 ms^{-1}

This is designed to be a discriminating question. The candidate's answer is wrong and does not make use of the earlier answers. At the equator, the force R exerted by the ground is given by

mg – R = ma
(Newton's second law)
R = 70 ×
(9.77 – 3.38 × 10⁻²)
R = 682 ≈ 680 N

The spinning of the Earth means that the reaction force R provided by the ground is reduced by an amount 'ma'.

Questions with model answers

A grade candidate – mark scored 8/9

 For help see Revise A2 Study Guide chapter 3.2

Examiner's Commentary

(1)   The diagram below shows an isolated positively charged metal sphere.

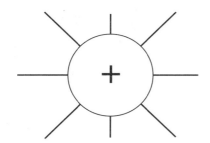

✔

Complete the diagram to show the electric field pattern around the charged sphere.   [2]

The candidate has lost one 'detail' mark. The radial field pattern is correct, but there is no indication of the direction of the field. The direction of the field is away from the positively charged sphere.

(2) (a)   Define electric potential at a point in space.   [2]

   Electric potential is equal to the work done per unit charge ✔
   for a charge brought from infinity to that point. ✔

   (b)   A metal sphere of radius 12 cm is suspended from the ceiling by means of a nylon thread. At a particular moment in time, the surface charge on the sphere is 3.2×10^{-8} C.

   (i)   Calculate the surface potential for the metal sphere.
         Data: $\varepsilon_0 = 8.85 \times 10^{-12}$ Fm^{-1} [3]

   $$V = \frac{Q}{4\pi\varepsilon_0 r}$$ ✔

   $$V = \frac{3.2 \times 10^{-8}}{(4\pi\varepsilon_0 \times 12 \times 10^{-2})}$$ ✔

   $$V = 2.4 \, kV$$ ✔

   (ii) Explain why the potential on the surface of the sphere rapidly decreases during a humid day.   [2]

   The charge on the sphere decreases. ✔
   Since charge $\propto$ potential, the potential starts to decrease as the charge is 'lost'.
   The charge is most likely to be lost through the surface moisture on the thread. ✔

The last answer is superb. The A grade candidate has provided a clear answer with the right blend of mathematics and physics.

Exam practice questions

A *Answers on pp. 62–64*

(1) **(a)** The diagram shows a positively charged sphere placed close to an earthed plate.

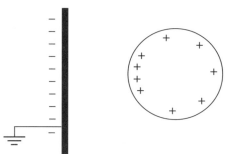

Draw the electric field pattern between the plate and the charged sphere. **[2]**

(b) The diagram shows a pair of parallel metal plates separated by a distance of 1.0 cm. The potential difference between the plates is 3.0 kV.

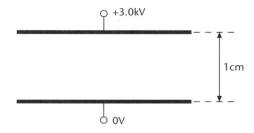

(i) On the diagram, draw a line representing the 1.0 kV **equipotential**. **[2]**

(ii) The electric field between the plates may be considered to be uniform. Calculate the field strength E between the plates. **[3]**

(iii) Calculate the force experienced by an electron between the plates due to the electric field of the plates.
Data: $e = 1.6 \times 10^{-19}$ C **[2]**

(2) Outline the similarities and the differences between electric and gravitational fields for **point** objects. **[4]**

(3) The diagram shows a 100μF capacitor connected in a circuit.

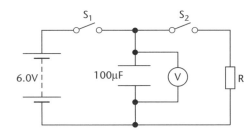

(a) With switch S_1 closed and S_2 open, state what happens to the potential difference (p.d.) across the capacitor. The capacitor is initially uncharged. **[1]**

(b) The switch S_1 is closed for a short period of time and then opened again. The graph below shows the variation of the p.d., V across the capacitor with time t when the switch S_2 is closed.

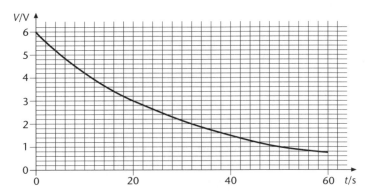

   (i) Calculate the energy stored by the fully charged capacitor. [3]

   (ii) Determine the time constant of the circuit. [1]

   (iii) Use your answer to **(b)(ii)** to calculate the resistance R of the resistor. [3]

   (iv) The p.d. across the capacitor decays exponentially. What does **exponential decay** mean? [1]

(4) The diagram below shows a capacitor designed from two identical sheets of aluminium foil and some paper.

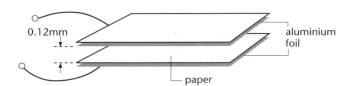

Each aluminium foil has a surface area of 8.0×10^{-3} m^2 and the thickness of the paper is 0.12 mm.

(a) Show that the capacitance C of the capacitor is equal to 1.4 nF.
   Data: $\varepsilon_0 = 8.85 \times 10^{-12}$ Fm^{-1}
   Relative permittivity of paper = 2.4 [2]

(b) The 1.4 nF capacitor is charged to a potential difference of 60 V.

   (i) Calculate the magnitude of the charge Q on one of the capacitor plates. [2]

(ii) The charged capacitor is then connected across another identical
1.4 nF uncharged capacitor via a switch S as shown in the diagram.

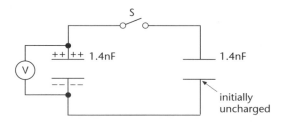

The switch S is closed.

1. Explain why the voltmeter reading decreases. **[2]**

2. Calculate the new reading on the voltmeter. **[3]**

Answers

(1) (a) Correct field pattern.
Correct field direction shown.

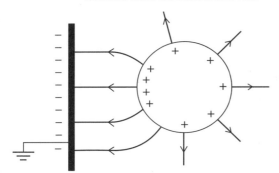

Examiner's tip

The radial field pattern of the charged sphere is distorted by the earthed plate. The electric field lines are normal to both the plate and the surface of the sphere.

(b)(i) Straight line between the plate.
This line lies at a distance of $\frac{1}{3}$ cm from the 0 V plate and curves in the region beyond the plates.

Examiner's tip

An equipotential is a line (or a surface) of equal potential. Since the field strength between the parallel plates is uniform, the potential V must change uniformly in the space between the plates. The 1.0 kV equipotential must therefore lie at a distance of $\frac{1}{3}$ cm from the negative plate. Beyond the parallel plates, the 1.0 kV equipotential is curved.

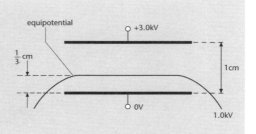

(ii) $E = -$ potential gradient

$$E = \frac{V}{d} \quad \text{(uniform field)}$$

$$E = \frac{3.0 \times 10^3}{0.01}$$

$$E = 3.0 \times 10^5 \text{ Vm}^{-1}$$

(iii) $F = Eq$
$F = 3.0 \times 10^5 \times 1.6 \times 10^{-19}$
$F = 4.8 \times 10^{-14}$ N

(2) Electric field is to do with **charges**.
Gravitational field is to do with objects having **mass**.
Both fields obey the inverse square law with distance.
Both fields are to do with 'force or *action* at a distance'.

Examiner's tip

There are a lot of other comments that could be added here. The points made above outline the main differences and similarities between the two fields.

(3) (a) The p.d. increases to 6.0 V.

(b)(i) $E = \frac{1}{2} V^2 C$
$E = \frac{1}{2} \times 6.0^2 \times 100 \times 10^{-6}$
$E = 1.8 \times 10^{-3}$ J

(ii) time constant $\approx$ 30 s

> ### Examiner's tip
> In a time equal to the time constant of the circuit, the p.d. decreases to a value of about 0.368 (which is e^{-1}) of its initial value. This value may be obtained from the graph by reading off the time when the p.d. is equal to
> $$0.368 \times 6.0 \approx 2.2 \text{ V}$$

(iii) time constant $= CR$
$CR \approx 30$ s
Therefore $R \approx \dfrac{30}{100 \times 10^{-6}}$
$R \approx 3.0 \times 10^5 \ \Omega$

(iv) In a given interval of time, the p.d. across the capacitor decreases by the same factor.

(4) (a) $C = \dfrac{\varepsilon_0 \, \varepsilon_r \, A}{d}$

$C = \dfrac{8.85 \times 10^{-12} \times 2.4 \times 8.0 \times 10^{-3}}{0.12 \times 10^{-3}}$

$C = 1.42 \times 10^{-9}$ F

> ### Examiner's tip
> The area A is the area of **overlap** between the two aluminium sheets. A common mistake made by candidates is to use the total area of the two capacitor plates.

(b)(i) $Q = VC$
$Q = 60 \times 1.42 \times 10^{-9}$
$Q = 8.52 \times 10^{-8} \approx 8.5 \times 10^{-8}$ C

> ### Examiner's tip
> The positive plate acquires a charge +Q and the negative plate a charge −Q. The total charge on both plates is zero. A capacitor is a device that separates charge. In the question, you are required to calculate the magnitude of the charge on one of the plates. This charge is given by the equation
> $$Q = VC$$

Answers

(ii) **1.** The charge on the original capacitor decreases as it is **shared** with the other capacitor.

Since $V \propto Q$, the p.d. across the capacitor decreases.

2. C_T = total capacitance

$C_T = C_1 + C_2 = 2.84$ nF

$V = \dfrac{Q}{C}$ (applied to the whole circuit)

$V = \dfrac{8.52 \times 10^{-8}}{2.84 \times 10^{-9}}$

$V = 30$ V

Examiner's tip

The total capacitance of the circuit is doubled. Since

$$V \propto \frac{1}{C}$$

*for the **same** total charge, the p.d. across the capacitors will be halved. Remember that the total charge in the circuit is conserved.*

Questions with model answers

C grade candidate – mark scored 6/10

 For help see Revise A2 Study Guide chapter 4.4

Examiner's Commentary

(1) Explain the terms in italics below, giving an example in each case.

    Antiparticle.   **[2]**

> *These are particles with the same mass but with*
> *opposite charge.* ✔

    Annihilation.   **[3]**

> *When particles collide and destroy each other.* ✗ (lack of detail)
>
> *The collision produces electromagnetic radiation.* ✔

(2) The diagram below illustrates the quark model for the proton.

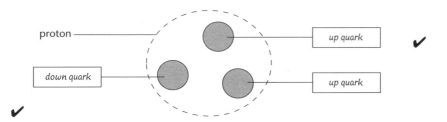

proton — up quark ✔

down quark — up quark

✔

Complete the diagram by labelling the type of quarks within the proton.   **[2]**

(3) (a) The radius *r* of the nucleus is given by

$$r = r_o A^{1/3}$$

where r_o is a constant and A is the nucleon number.

Use this expression to determine the ratio   **[2]**

$$\frac{\text{radius of } ^{235}\text{U nucleus}}{\text{radius of } ^{12}\text{C nucleus}}$$

$$\text{ratio} = \frac{\sqrt[3]{235}}{\sqrt[3]{12}}$$ ✔

$$\text{ratio} = \frac{6.17}{2.29} = 2.7$$ ✔

(b) Suggest why the density of nuclear material is very much greater than the density of ordinary matter, like a piece of rock.   **[1]**

> *A rock is made from atoms. The radius of the atom is*
> *larger than the nucleus.* ✗

*The candidate has had success with defining the first term, but there is no named antiparticle. As a result, the candidate has lost one mark. A good example of an antiparticle is the positron, which has the same mass as that of an electron, but with a positive charge of $+1.6 \times 10^{-19}$ C. The candidate's answer for annihilation lacks detail. There is no mention of an interaction between **matter** and **antimatter**. When a particle encounters its antiparticle, the chances are that they will 'destroy' each other. According to Einstein's equation*

$$\Delta E = \Delta mc^2$$

the total mass of both particles is converted into electromagnetic radiation in the form of γ-ray photons.

A proton consists of two up quarks and a single down quark. The up quark has a charge $+\frac{2}{3}e$ and the down quark $-\frac{1}{3}e$, making a total charge of $+e$ for the proton.

*There is only a mark for this part of the question. The candidate's answer lacks completeness. The candidate has failed to realise that roughly the **same** amount of mass is contained in a much **smaller** volume of the nucleus – the electrons are not massive compared to a proton or a neutron, therefore their contribution to the mass of the atom is negligible. Hence the density of the nucleus is much greater than that of the atoms. Incidentally, the radius of most atoms is ~ 10^{-10} m, whereas the radius of most nuclei is ~ 10^{-15} m.*

Questions with model answers

A grade candidate – mark scored 12/12

? *For help see Revise AS Study Guide chapter 3.4*

(1) Complete the table below by naming the remaining three fundamental interactions (forces) and the exchange particles associated with the interactions. **[6]**

interaction	exchange particle	
gravitational	graviton	
electromagnetic	photons	✔ ✔
strong nuclear force	gluons	✔ ✔
weak nuclear force	W^+, W^- and Z particles	✔ ✔

Examiner's Commentary

The weak nuclear force is mediated through the particles listed above by the candidate. These particles are collectively referred to as the Intermediate Vector Bosons.

(2) Distinguish between leptons and hadrons. **[4]**

Hadrons are made up of quarks. ✔
Hadrons are effected by the strong nuclear force. ✔
Leptons are fundamental particles. ✔
Leptons feel the weak nuclear force. ✔

(3) The Feymann diagram below illustrates the decay of a neutron.

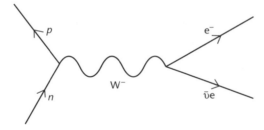

With reference to the diagram, explain each stage of the decay. **[2]**

A neutron decays into a proton and the W^- boson. ✔
In time, the W^- boson decays into an electron and an anti-neutrino. ✔
The decay shown above is that for β-decay within a nucleus.

Exam practice questions

 Answers on pp. 69–71

(1) **(a)** The mean radius of the ^{4}He nucleus is 1.9×10^{-15} m. Calculate the mean density of the helium-4 nucleus. You may assume that a neutron has the same mass as a proton.
Data: mass of proton $\approx 1.7 \times 10^{-27}$ kg. **[3]**

(b) Explain why the mean density of any nucleus is independent of its mass. **[3]**

(c) The mean separation of the protons within the nucleus of helium–4 is about 3.8×10^{-15} m.

   (i) Calculate the gravitational force, F_G between the two protons.
   Data: $G = 6.67 \times 10^{-11}$ Nm^{2}kg^{-2} **[2]**

   (ii) Calculate the electrical force, F_E between the two protons.
   Data: $\varepsilon_0 = 8.85 \times 10^{-12}$ Fm^{-1}
   $e = 1.6 \times 10^{-19}$ C. **[2]**

   (iii) Explain why the answers to **(c)(i)** and **(c)(ii)** suggests that there must be another force acting on the protons when inside the nucleus. **[2]**

(2) Nuclei, like atoms, absorb and emit electromagnetic radiation in the form of photons. For a single nucleus of deuterium ($^2_1$H), a minimum energy of 2.2 MeV is required to **just** separate all the nucleons.

(a) Explain what is meant by

   (i) a nucleon, **[1]**

   (ii) a photon. **[1]**

(b) Calculate the wavelength of the electromagnetic radiation absorbed by the nucleus of deuterium to just free all the nucleons.
Data: 1 eV = 1.6×10^{-19} J
   $c = 3.0 \times 10^8$ ms^{-1}
   $h = 6.63 \times 10^{-34}$ Js **[3]**

(3) The Sun produces energy by means of fusion reactions. One such reaction is
$$^1\text{H} + {}^1\text{H} \rightarrow {}^2\text{H} + e^+ + \upsilon$$

(a) Identify the particles ^{1}H and e^+. **[2]**

(b) Explain how the reaction above releases energy. **[2]**

Exam practice questions

(c) Explain why very high temperatures are necessary for the above reaction to occur. **[2]**

(d) The graph below shows the variation of the average binding energy per nucleon (B.E./nucleon) with the nucleon number A.

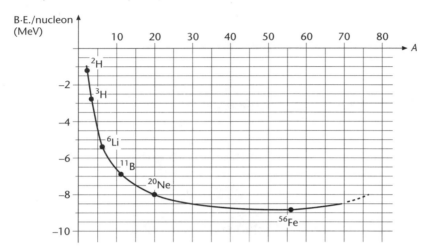

(i) Explain why ^1H does not appear on the graph. **[1]**

(ii) Use the graph to determine the energy released in the reaction.
Data: 1 eV = 1.6×10^{-19} J **[2]**

(4) One possible fission reaction within a nuclear reactor is shown below.

$$^{235}U + {}^1n \rightarrow \boxed{} \rightarrow {}^{96}Rb + {}^{138}Cs + X{}^1n$$

(a) Complete the reaction above by inserting the appropriate symbols in the box. **[1]**

(b) Identify the number X of neutrons released in the reaction above. **[1]**

(c) Calculate the energy released in the above reaction.
Data: mass of ^{235}U nucleus = 3.90×10^{-25} kg
mass of ^{96}Rb nucleus = 1.59×10^{-25} kg
mass of ^{138}Cs nucleus = 2.29×10^{-25} kg
mass of ^1n = 1.67×10^{-27} kg
c = 3.00×10^8 ms^{-1} **[4]**

(d) Determine the energy released by 1 kg of pure uranium-235 fuel. **[3]**

(e) State one major disadvantage of using nuclear fuel. **[1]**

Answers

(1) (a) $\rho = \dfrac{M}{V}$

$M = 4 \times 1.7 \times 10^{-27} = 6.8 \times 10^{-27}$ kg
$V = \frac{4}{3}\pi r^3 = \frac{4}{3}\pi \times (1.9 \times 10^{-15})^3$
$V = 2.873 \times 10^{-44}$ m^3

$\rho = \dfrac{6.8 \times 10^{-27}}{2.873 \times 10^{-44}}$

$\therefore \rho = 2.37 \times 10^{17} \approx 2.4 \times 10^{17}$ kgm^{-3}

(b) Mass $\propto A$
$r = r_o A^{1/3}$ and $V = \frac{4}{3}\pi r^3$
$\therefore$ volume $\propto A$
density $= \dfrac{\text{mass}}{\text{volume}}$
The density is independent of the nucleon number A.

> **Examiner's tip**
>
> *The mean density of the nucleus is the same because it consists of the same matter.*

(c)(i) $F_G = \dfrac{GMm}{r^2}$

$F_G = \dfrac{6.67 \times 10^{-11} \times (1.7 \times 10^{-27})^2}{(3.8 \times 10^{-15})^2}$

$F_G = 1.33 \times 10^{-35} \approx 1.3 \times 10^{-35}$ N

(ii) $F_E = \dfrac{Qq}{4\pi\varepsilon_0 r^2}$

$F_E = \dfrac{(1.6 \times 10^{-19})^2}{4\pi\varepsilon_0 (3.8 \times 10^{-15})^2}$

$F_E = 15.9 \approx 16$ N

(iii) The gravitational force is weak but attractive and the electrical force is much larger but repulsive.
With the forces above, there is a **net repulsive** force.
There must therefore be another attractive force holding the protons.

> **Examiner's tip**
>
> *The nucleons are held within the nucleus by the strong nuclear interaction. This interaction exists between the quarks that make up the nucleons.*

(2) (a)(i) nucleon $\rightarrow$ a proton or a neutron.

(ii) photon $\rightarrow$ quantum (or packet) of electromagnetic energy.

Answers

(b) E = energy of photon
$E = 2.2$ MeV $= 2.2 \times 10^6 \times 1.6 \times 10^{-19} = 3.52 \times 10^{-13}$ J
$E = hf$ and $c = f\lambda$
$E = \dfrac{hc}{\lambda}$

$\lambda = \dfrac{6.63 \times 10^{-34} \times 3.0 \times 10^8}{3.52 \times 10^{-13}}$

$\lambda = 5.65 \times 10^{-13} \approx 5.7 \times 10^{-13}$ m

Examiner's tip

The photons are energetic enough to split the nucleus. The photons belong to the γ-ray region of the electromagnetic spectrum. The other method for separating the nucleons would be to use energetic particles in the form of electrons and protons.

(3) (a) ^{1}H is a proton.
e^+ is a positron (the anti-particle of the electron).

(b)   In the reaction, mass **decreases**.
According to Einstein's equation, $\Delta E = \Delta mc^2$
a decrease in mass means that energy is released
in the reaction.

Examiner's tip

Some of the mass in this reaction is converted into energy. This is in accordance with Einstein's equation mentioned above.

(c)   The protons are positively charged, they therefore **repel** each other.
At high temperature, the protons move faster and therefore the chance
of overcoming the repulsive force is greater.

Examiner's tip

The protons may be assumed to behave like the molecules of an ideal gas. Therefore, the mean kinetic energy E_k of the protons is given by
$$E_k = \tfrac{3}{2} kT$$
As the temperature increases, E_k increases. For the protons to 'fuse' together, they must be close enough to be influenced by the strong nuclear force. This force is very short-ranged. The kinetic energy of the protons must be greater than the electrical potential energy between the protons. This happens when the temperature is about 10^8 K.

(d)(i) ^{1}H is a proton.
It has no binding energy because there are no other nucleons.

(ii)   B.E. / nucleon ≈ -1.2 MeV
E = energy **released**
$E \approx 2 \times (1.2 \times 10^6 \times 1.6 \times 10^{-19})$
$E \approx 3.8 \times 10^{-13}$ J

(4) (a) ^{236}U label inserted within the box.

(b) The nucleon number must be conserved.
There are two neutrons released in the reaction.

(c) Δm = mass defect
$\Delta m = (1.59 \times 10^{-25} + 2.29 \times 10^{-25} + 2 \times 1.67 \times 10^{-27})$
$\quad - (3.90 \times 10^{-25} + 1.67 \times 10^{-27})$
$\Delta m = -3.30 \times 10^{-28}$ kg
$\Delta E = \Delta mc^2$
$\Delta E = -3.30 \times 10^{-28} \times (3.0 \times 10^8)^2$
$\Delta E = -2.97 \times 10^{-11}$ J

(d) Number of nuclei = $\dfrac{1 \text{ kg}}{\text{mass of uranium-235 nucleus}}$

Number of nuclei = $\dfrac{1}{3.90 \times 10^{-25}} = 2.5641 \times 10^{24}$

Energy released = $2.5641 \times 10^{24} \times 2.97 \times 10^{-11}$
Energy released = 7.615×10^{13} J $\approx 7.62 \times 10^{13}$ J

(e) Radioactive waste from nuclear fuel lasts for a very long time.

Questions with model answers

C grade candidate – mark scored 8/12

 For help see Revise A2 Study
Guide chapters 6.2 and 6.3

Examiner's Commentary

(1) State the three assumptions, known collectively as the cosmological principle.   **[3]**

> The matter in the universe is spread out evenly
> and it is not 'lumpy'.   ✔
> The universe is the same in all directions.   ✔
> The laws of physics are universal.   ✔

*This is a good start, with the candidate recalling all three fundamental assumptions in cosmology. When we look at the universe in detail, then we do find matter 'lumped' together in the form of solar systems, galaxies etc., but overall, the universe appears to have the **same** mean density in all directions. The third statement represents an interesting assumption that the laws of physics, as applicable on the Earth, have been true since the birth of the universe.*

(2) State Hubble's law and use it to estimate the age of the universe. Data: Hubble constant, $H_0 \approx 3.3 \times 10^{-18}$ s^{-1} **[4]**

> $$age = \frac{1}{H_0}$$ ✔
>
> $$age \approx \frac{1}{3.3 \times 10^{-18}} \approx 3.0 \times 10^{17} \text{ s}$$ ✔

There are four marks in total and the candidate has managed to secure only two. The candidate has not scrutinised the question because there is no mention of Hubble's law. Hubble's law gives a relationship between the speed v of recession of a star or galaxy and its distance d from us. It may be written as: $v \propto d$ or $v = H_0 d$ where H_0 is the Hubble constant.

(3) Explain what is meant by an 'open' universe.   **[2]**

> The universe will expand forever.   ✔
> The reason for this is the speed of the galaxies
> is very large.   ✗

The reason why the universe may expand indefinitely is because the gravitational attraction between the matter in the universe is not strong enough to halt the expansion of the galaxies that resulted from the Big Bang.

(4) For a 'closed' universe, the mean density of matter within the universe must be greater than 1.9×10^{-26} kgm^{-3}. Current evidence suggests that the mean density is about three protons in every cubic metre of space. Discuss what consequence this may have on the fate of the universe. Data: mass of proton = 1.7×10^{-27} kg.   **[3]**

> The universe will expand forever because the density is
> less than 1.9×10^{-26} kgm^{-3}.   ✔
> Current density is only
> $3 \times 1.7 \times 10^{-27} = 5.1 \times 10^{-27}$ kgm^{-3}.   ✔

The answer above is brief. At present, there is some dispute that the actual density of the universe is much greater than current measurements because of the existence of the mysterious 'dark matter'. For some discussion of this, or other relevant points, the candidate may have received some credit.

A grade candidate – mark scored 6/6

For help see Revise A2 Study Guide chapter 5.2

Examiner's Commentary

(1) The diagram shows the path of two rays from an object which are incident on the cornea of an eye.

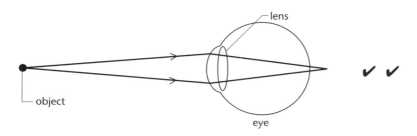

✔ ✔

(a) On the diagram, draw the paths of the rays for a person who is suffering from long sight. [2]

(b) Name the corrective lens required to correct the defect shown in (a). [1]

A convex lens may be used to correct the defect. ✔

(2) A person suffering from hypermetropia has a near point 1.10 m from the eyes.

Calculate the power P of the corrective lens needed to bring the person's near point to 0.25 m from the eyes. [3]

Power $P = \dfrac{1}{f}$ where f = focal length of the lens. ✔

$\dfrac{1}{f} = \dfrac{1}{u} + \dfrac{1}{v}$ $u = +0.25m$ and $v = -1.10m$ ✔

$\dfrac{1}{f} = \dfrac{1}{0.25} - \dfrac{1}{1.10} = 3.09$

Therefore $P = 3.1$ D ✔

This is a perfect answer. The calculations are easy to decipher and the correct unit for power has been given. This is a typical answer from an A grade candidate.

Exam practice questions

A *Answers on pp. 76–77*

(1) (a) What evidence is there that the universe is expanding? **[2]**

(b) The diagram shows the position of a spectral line from hydrogen as observed in the laboratory on the Earth and that from a star in a distant galaxy.

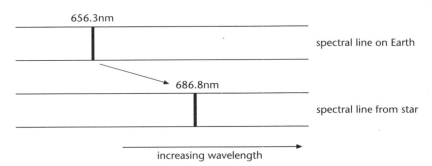

(i) State why the wavelength of light from the star is different from that observed on the Earth. **[1]**

(ii) For a spectral line of wavelength λ from a stationary source, the change in wavelength $\Delta\lambda$ when the source is moving at a speed v is given by the Doppler equation

$$\frac{\Delta\lambda}{\lambda} = \frac{v}{c}$$

where c is the speed of light in a vacuum.
Calculate the speed of the star as observed from the Earth.
Data: $c = 3.00 \times 10^8 \text{ ms}^{-1}$ **[2]**

(2) This question is about our Solar System and Kepler's third law.

(a) Show that the period T of a planet orbiting our Sun at a mean distance r away from it is given by

$$T^2 = \frac{4\pi^2}{GM} r^3$$

where M is the mass of the Sun. The equation given above is often referred to as Kepler's third law. **[3]**

(b) The table below shows information about some of the planets in our Solar System.

Planet	T / years	r / A.U.
Mercury	0.24	0.39
Venus	0.61	0.72
Earth	1.00	1.00
Mars	1.88	1.52
Jupiter	11.86	5.20

1 A.U. = mean distance between the Sun and the Earth.

Use the information given in the table to show that Kepler's third law applies to the planets in our Solar System. **[3]**

(c) Our Sun, together with its Solar System, is moving through space because our galaxy is rotating. The mean distance of our Sun from the galactic centre is 25 000 light-years and its orbital period is 2.0×10^8 years. Use Kepler's third law to calculate the mass M of our galaxy.
Data: $G = 6.67 \times 10^{-11}$ Nm^2kg^{-2}
  $c = 3.0 \times 10^8$ ms^{-1} **[3]**

(3) **(a)** On the axes below, sketch the variation with frequency f of the minimum intensity I of sound detectable by a person with normal hearing.
Useful data: Ear is most sensitive at 4 kHz.
Threshold intensity, $f_0 \approx 1 \times 10^{-12}$ Wm^{-2}

[3]

(b) In a factory, Health Officers record sound intensity to be 4.1×10^{-2} Wm^{-2}. It is recommended that workers in the factory wear ear guards so as to reduce the sound incident on the ears to 3.0×10^{-5} Wm^{-2}.
Calculate the reduction in sound intensity, in dB. **[4]**

(4) **(a)** Describe some of the effects of ionising radiation on human beings. **[2]**

(b) Radioisotopes have many uses in non-invasive techniques of diagnosis. Discuss the use of a specific tracer in medicine. **[5]**

Answers

(1) (a)   Distant stars and galaxies are moving away from us.
Light from such sources is red-shifted.

Examiner's tip

Red-shift means that the measured wavelength of light from moving sources is longer than if the source was stationary.

(b)(i) The star is receding from us, therefore the light from it is red-shifted.

(ii) $\dfrac{\Delta\lambda}{\lambda} = \dfrac{(686.8 - 656.3)}{656.3}$

$\dfrac{\Delta\lambda}{\lambda} = 0.04647$

$v = 0.04647 \times 3.00 \times 10^8$
$v \approx 1.39 \times 10^7\ \text{ms}^{-1}$

(2) (a) $F = \dfrac{GMm}{r^2}$ $F = m(\omega^2 r)$ and $\omega = \dfrac{2\pi}{T}$

Therefore $\dfrac{GM}{r^2} = \omega^2 r$

$\dfrac{GM}{r^3} = \dfrac{4\pi^2}{T^2}$

$T^2 = \dfrac{4\pi^2}{GM}\,r^3$

Examiner's tip

This proof is worth remembering. It uses all the key ideas from Newtonian mechanics.

(b)   If $T^2 \propto r^3$, then $\dfrac{T^2}{r^3}$ must be a constant.

The values for $\dfrac{T^2}{r^3}$ are

0.971, 0.997, 1.000, 1.006 and 1.000
for Mercury, Venus, Earth, Mars and Jupiter respectively.

Hence Kepler's third law is applicable to our Solar System.

(c) $T^2 = \dfrac{4\pi^2}{GM}\, r^3$

where $r = 25\,000 \times 3.0 \times 10^8 \times 365 \times 24 \times 3600 = 2.365 \times 10^{20}$ m
and $T = 2.0 \times 10^8 \times 365 \times 24 \times 3600 = 6.307 \times 10^{15}$ s

$M = \dfrac{4\pi^2 r^3}{G\, T^2}$

$M = \dfrac{4\pi^2 \times (2.365 \times 10^{20})^3}{6.67 \times 10^{-11} \times (6.307 \times 10^{15})^2}$

$M = 1.96 \times 10^{41}$ kg $\approx 2.0 \times 10^{41}$ kg   (Equivalent to $\sim 10^{11}$ solar masses)

Examiner's tip

It is vital that the distance and the period are calculated in metres and seconds respectively. If this is not done, then most likely you would lose marks since the final mass will not be in kilograms. This is an 'extension' question where a specific model is applied to another situation. In this case, Kepler's third law which has been shown to hold true for the planets in our Solar System is now being applied to the entire galaxy. The mass M is in fact the mass of the galaxy contained within the orbit of our Sun. Since the majority of the mass is contained within the galactic centre, the answer above is not too far from the correct value.

(3) (a)   The minimum ought to have the following values:

$\log_{10} f = \log_{10} 4000 \approx 3.6$

and $\log_{10} I = \log_{10} (1 \times 10^{-12}) = -12$

Correct shape of the graph.

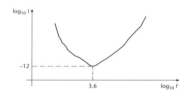

(b)   Intensity level $= 10 \log_{10} \left(\dfrac{I}{I_0} \right)$

Original value $= 10 \log_{10} \left(\dfrac{4.1 \times 10^{-2}}{10^{-12}} \right) = 106.1$ dB

Final value $= 10 \log_{10} \left(\dfrac{3.0 \times 10^{-5}}{10^{-12}} \right) = 74.8$ dB

Reduction $= 106.1 - 74.8 = 31.3$ dB
Reduction ≈ 31 dB.

(4) (a)   Any *two* from:
Genetic mutation of cells,
Cells may be destroyed
leading to serious illness or death.

(b)   A tracer is a radioactive substance introduced into the body of a patient.
It is used for detecting abnormalities in the function of organs.
Commonly used tracer is **technetium** which emits γ-rays and has a short half-life of 6 hours.
The half-life is short enough, so little damage is done to living cells.
The γ-ray photons have energy of 0.14 MeV and as such are easier to detect.

Examiner's tip

The chances are that the examiners may also assess the candidate's 'Quality of Written Communication' in a question like (b). It is therefore very important that extra care is given to sentence construction, spelling and grammar.

Mock Exam

Time: 1 hour 20 minutes   Maximum marks: 80

Instructions
Answer **all** questions in the spaces provided.
Show all steps in your working.
The marks allocated for each question are shown in brackets.
Any data required for a question is given where appropriate.

Grading
Boundary for A grade   64/80
Boundary for C grade   48/80

---

(1) The drag force F acting on a car travelling at a constant speed v is given
by the equation

$$F = k\,v^2$$

where k is a constant for a given car.

(a) Determine the unit for k and suggest one factor related to the car that may
affect the magnitude of k.

....

....

.... **[2]**

(b) For the car travelling at a constant velocity of 27 ms^{-1}, the drag force
is 670 N.
(i) State and explain the magnitude of the motive force produced by the
car engine.

....

.... **[2]**

(ii) Calculate the value for k.

....

.... **[2]**

(iii) Calculate the power developed by the engine when the car is travelling
at a constant speed of 27 ms^{-1}.

....

.... **[2]**

(iv) Show that the motive power P developed by the engine is given by
$$P \propto v^3$$
Hence determine the change in the motive power of the engine when the speed of the car is <u>reduced</u> by 20%. Suggest one advantage of reducing car speed.

....

....

.... **[3]**

[11 marks]

(2) (a) The diagram shows a current-carrying conductor placed between the poles of a permanent magnet.

(i) Explain why the current-carrying conductor experiences a force.

....

.... **[2]**

(ii) Describe Fleming's left hand rule. Use this rule to predict the direction in which the conductor would move.

....

....

.... **[3]**

(iii) A length of 3.2 cm of the conductor lies in the uniform field of the magnet. When carrying a current of 5.0 A, the force experienced by this length of the conductor is 6.1×10^{-3} N. Determine the magnetic flux density (magnetic field strength) of the magnetic field.

....

....

.... **[3]**

(b) The diagram shows a cable carrying current placed in a uniform magnetic field of flux density B.

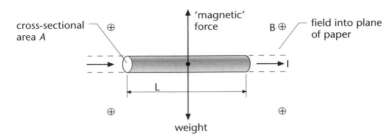

The current in the cable is such that the force on the cable created by the magnetic field is sufficient to support the weight of the cable itself. Show that the current I is given by

$$I = \frac{\rho A g}{B}$$

where ρ is the density of the material of the cable, A is the cross-sectional area of the cable and g is the gravitational field strength.

....

....

....

....

.... **[3]**

[11 marks]

(3) The graph below shows the current-voltage characteristic of a lamp provided by a manufacturer.

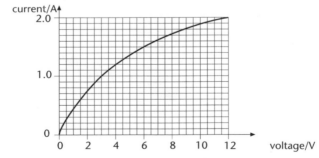

(a) State how the resistance of the lamp changes as the voltage across it is increased.

....

.... **[1]**

(b) A student buys two identical lamps from the manufacturer. The lamps are connected to a 12 V d.c. supply having negligible internal resistance. Determine the total resistance of the circuit when both lamps are connected to the supply

  (i) in a series combination,

  ....

  .... **[2]**

  (ii) in a parallel combination.

  ....

  .... **[2]**

(c) The diagram shows **one** of the lamps and a 10 Ω resistor connected in series to a battery.

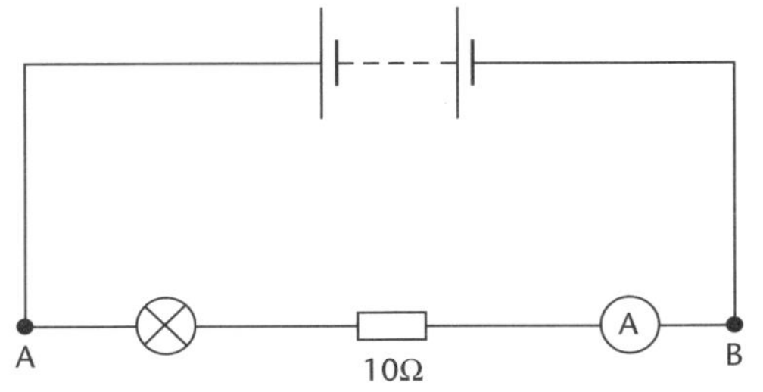

The current measured by the ammeter is 1.0 A. The ammeter has negligible resistance. Calculate the potential difference across A and B.

....

....

.... **[5]**

[10 marks]

(4) A metal wire of cross-sectional area 3.2×10^{-8} m^2 has a length 2.5 m. At room temperature, the wire has a resistance of 4.6 Ω.
(a) Calculate the resistivity of the material the wire is made from.

....

.... **[3]**

(b) The diagram shows the wire wound into a tight coil and connected to a supply of negligible internal resistance.

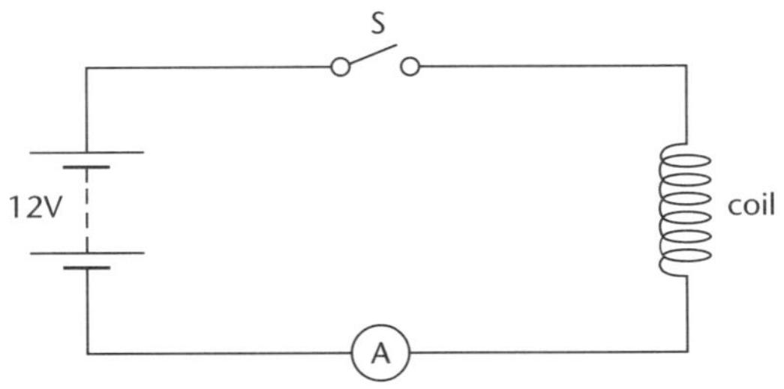

When the switch S is closed, the coil eventually gets 'red hot'. The graph below shows the variation of the circuit current I with time t.

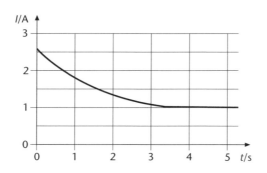

(i) Show why the current at $t = 0$ s is 2.6 A. Explain your answer.

....

.... **[2]**

(ii) With reference to the movement of electrons within the material of the wire, explain the shape of the graph.

....

....

.... **[3]**

[8 marks]

(5) (a) Explain what is meant by electromagnetism.

....

.... **[1]**

(b) Describe and explain an experiment, with the aid of a diagram if necessary, that may be done in the laboratory to show the phenomenon of electromagnetism.

....

....

....

....

.... **[4]**

[5 marks]

(6) (a) State three assumptions of the kinetic theory of gases.

....

....

.... [3]

(b) State the meaning of the symbols in the equation
$$pV = \tfrac{1}{3} Nm\langle c^2 \rangle.$$

....

....

.... [5]

(c) Air consists principally of oxygen and nitrogen. Both gases may be considered to behave as ideal gases at room temperature of 20°C.
  (i) Show that the mean translational kinetic energy E_k of molecules of an ideal gas is given by
$$E_k = \tfrac{3}{2} kT$$
  where k is the Boltzmann constant.

....

....

.... [3]

  (ii) Calculate the mean translational kinetic energy E_k of nitrogen molecules at 20°C.
  Data: $k = 1.38 \times 10^{-23}$ JK^{-1}

....

.... [2]

  (iii) Oxygen molecules are more massive than nitrogen molecules. State and explain how the answer to (c)(ii) would change, if at all, for oxygen molecules.

....

.... [2]

(iv) Explain what is meant by the **internal energy** of a gas. Calculate the total internal energy of one mole of nitrogen.
Data: Avogadro number, $N_A = 6.02 \times 10^{23}$ mol^{-1}

....

....

.... **[3]**

[18 marks]

(7) **(a)** Explain what is meant by the **binding energy** of a nucleus.

....

....

.... **[1]**

(b) The graph below shows the average binding energy per nucleon
(B.E. / nucleon) against the nucleon number A.

(i) What is the significance of the average binding energy per nucleon being negative?

....

.... **[1]**

(ii) Use the graph to determine the binding energy of a single nucleus of ^{56}Fe.

....

.... **[2]**

(c) One of the naturally occurring isotopes of radium is radium-226. The nuclei of radium-226 emit alpha (α) particles.

(i) Complete the following nuclear decay equation for a single nucleus of radium-226.

$$^{226}_{88}\text{Ra} \rightarrow \text{Rn} + \text{He}$$

[2]

(ii) Calculate the energy released by the decay of a single nucleus of radium-226.

Data: 1 u = 1.66×10^{-27} kg

$c = 3.00 \times 10^8$ ms^{-1}

mass of alpha particle = 4.00 u

mass of Ra nucleus = 225.98 u

mass of Rn nucleus = 221.97 u

....

....

.... [4]

(iii) State in what form the energy in **(c)(ii)** is released.

.... [1]

(iv) The half-life of radium-226 is about 1600 years. Suggest one reason why radium cannot be used as a viable source of energy.

....

.... [1]

[12 marks]

(8) According to a student:

'The only factor affecting the power of a radioactive sample emitting beta-particles is its activity and the power remains constant over a long period of time.'

Explain what is meant by **activity**. Discuss whether the comment made by the student is correct or not.

....

....

....

....

.... [5]

[5 marks]

Answers

(1) (a) $k = \dfrac{F}{v^2}$

Therefore k has units Ns^2m^{-2} .

Drag would depend on the frontal area of the car.

(b)(i) The car is travelling at constant velocity and therefore has no acceleration.
According to $F = ma$, the **net** force on the car will be zero.
Hence, the force provided by the engine, the motive force = 670 N.

(ii) $k = \dfrac{F}{v^2}$

$k = \dfrac{670}{27^2}$

$k = 9.19 \times 10^{-1} \approx 9.2 \times 10^{-1}\ Ns^2m^{-2}$

(iii) Power $= Fv$
$P = 670 \times 27$
$P = 1.81 \times 10^4 \approx 1.8 \times 10^4\ W$

(iv) $P = Fv$
$P = (kv^2)v = kv^3$
k is a constant, therefore $P \propto v^3$

With a 20% reduction in speed, the power will decrease to $(0.80)^3$ of its original value. The engine power will therefore be 51% of its previous value. (There is change in power of 49%.)

One possible advantage would be using less fuel and therefore reduced environmental pollution.

Examiner's tip

You can do the question by calculating the drag force on the car at the new speed using $F = kv^2$ and then applying $P = Fv$. The route shown above is concise and shows that the actual value of k does not matter in this case. It is also advantageous to use the information provided by the examiners. In this case, the relationship $P \propto v^3$.

(2) (a)(i) The conductor is surrounded by a magnetic field.
This field of the conductor 'interacts' with that of the magnet and hence the conductor moves.

(ii) **F**irst finger points in the direction of the magnetic **f**ield.
Se**c**ond finger points in the direction of the **c**urrent.
Thu**m**b gives the direction of the **m**otion or force.
The thumb and fingers are at right angles to each other.

The wire deflects in the (positive) z-direction.

(iii) $F = BIL$
$B = \dfrac{F}{IL} = \dfrac{6.1 \times 10^{-3}}{(5.0 \times 3.2 \times 10^{-2})}$
$B = 3.81 \times 10^{-2} \approx 3.8 \times 10^{-2}\ T$

(b)   weight = 'magnetic' force

$Mg = BIL$

$(\rho V)g = BIL$

$(\rho AL)g = BIL$

Hence $I = \dfrac{\rho Ag}{B}$

(3) (a)   As the p.d. increases, the resistance of the lamp increases.

> ### Examiner's tip
>
> *It is very sad when candidates guess the answer here. You can play safe and calculate the value of the resistance at low and high voltages **using** the graph.*

(b)(i)   The bulbs are identical, therefore the p.d. across each is 6.0 V

From the graph, current = 1.5 A

$R = \dfrac{V}{I} = \dfrac{6.0}{1.5} = 4.0 \ \Omega$

Total resistance, $R_T = 4.0 + 4.0 = 8.0 \ \Omega$

(ii)   Since the bulbs are connected in parallel, the p.d. across each lamp is 12 V

From the graph, current = 2.0 A

$R = \dfrac{V}{I} = \dfrac{12}{2.0} = 6.0 \ \Omega$

$R_T = \dfrac{R_1 R_2}{(R_1 + R_2)}$

> ### Examiner's tip
>
> *This question cannot be done without the graph given at the start of the question. Use all the information given by the examiners.*

$R_T = \dfrac{6.0 \times 6.0}{12.0}$

$R_T = 3.0 \ \Omega$

(c)   From the graph, when current is 1.0 A the potential difference across the bulb is 3.0 V.

$R_{BULB} = \dfrac{V}{I} = \dfrac{3.0}{1.0} = 3.0 \ \Omega$

$R_T = R_1 + R_2 = 10 + 3.0$

$R_T = 13 \ \Omega$

$V = IR$

$V = 1.0 \times 13$

$V = 13 \ V$

(4) (a) $R = \dfrac{\rho \ell}{A}$

$\rho = \dfrac{RA}{\ell} = \dfrac{4.6 \times 3.2 \times 10^{-8}}{2.5}$

$\rho = 5.89 \times 10^{-8} \approx 5.9 \times 10^{-8} \ \Omega m$

(b)(i)   When the switch is closed, the coil is at room temperature and therefore has resistance of 4.6 Ω

$I = \dfrac{V}{R} = \dfrac{12}{4.6} = 2.6 \ A$

(ii) The wire starts to heat up as current passes.
This leads to greater vibration of the ions and the path of the electrons
is opposed more. The resistance of the wire therefore increases,
which is shown by the decreasing current.
Eventually, the coil attains a constant temperature,
therefore the current is constant.

Examiner's tip

*The examiners are looking for at least three distinct comments from the candidate.
It is important that the candidate has something to write about the movement of the
electrons within the wire.*

(5) (a) A current-carrying conductor is surrounded by a magnetic field.

(b) A current carrying conductor is used.
A plotting compass is used.
This deflects in different directions when placed around the conductor.
Labelled diagram of the apparatus.

Examiner's tip

*There are many other ways of showing that a current carrying conductor is surrounded by a
magnetic field. A description of an electromagnet picking up paper clips or iron filings would
be worthy of credit. Always keep an eye on the marks available. There are four marks
reserved for the description of the experiment, so resist making a single bold statement.*

(6) (a) Any *three* from:
The molecules have a negligible volume compared with the volume of the container.
The molecules collide elastically with each other (or the container walls).
The forces between the molecules are negligible, except during collisions.
The motion of the molecules is random.
The time for molecular collisions is negligible compared to the time
between the collisions.
There are a very large number of molecules inside the container.

(b) p is the pressure exerted by the gas.
V is the volume of the gas (or the container).
N is the number of molecules.
m is the mass of each gas molecule.
$\langle c^2 \rangle$ is the mean square speed of the molecules.

(c)(i) $pV = nRT$
$nRT = \frac{1}{3} Nm \langle c^2 \rangle$
For one mole of gas, $n = 1$ and $N = N_A$, the Avogadro constant.
$RT = \frac{1}{3} N_A m \langle c^2 \rangle$

$$\frac{1}{3} m \langle c^2 \rangle = \left(\frac{R}{N_A} \right) T$$

But $k = \dfrac{R}{N_A}$

$\therefore \frac{1}{2} m \langle c^2 \rangle = \frac{3}{2} kT$
$\frac{1}{2} m \langle c^2 \rangle$ is the mean translational kinetic energy, E_k of the molecules.

(ii) $E_k = \frac{3}{2} \times 1.38 \times 10^{-23} \times (273 + 20)$
$E_k = 6.07 \times 10^{-21} \approx 6.1 \times 10^{-21}$ J

Examiner's tip

It is vital that the temperature is written in kelvin. T is the thermodynamic temperature.

(iii) The mean translational kinetic energy for oxygen will be the same. This is because $E_k \propto T$, hence independent of the mass.

Examiner's tip

At the same temperature T, the mean translational kinetic energy E_k for all types of molecules is the same. This is because $E_k \propto T$. The more massive particles will move slowly, however,

$$\frac{1}{2} m \langle c^2 \rangle$$

*will be the **same** for all molecules.*

(iv) Internal energy = sum of K.E. and P.E. of molecules.
E = energy of one mole of gas
$E = 6.07 \times 10^{-21} \times 6.02 \times 10^{23}$ (Assume P.E. = 0)
$E = 3.65 \times 10^3 \approx 3.7 \times 10^3$ J

(7) (a) The binding energy of a nucleus represents the energy required to separate the nucleus into its individual nucleons.

(b)(i) The B.E. per nucleon is **negative** because it implies that external energy must be supplied to the nucleus in order to separate the nucleons.

Examiner's tip

*Within the nucleus, the nucleons are held together by the **strong nuclear force**. To move the nucleons apart, requires external energy. Hence the initial energy of the nucleus must be negative. If the energy was assigned a positive sign, then this would imply that the nucleons have kinetic energy and are already free from the effect of the strong interaction.*

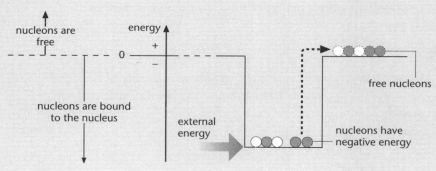

(ii) B.E. = number of nucleons × B.E. /nucleon
B.E. ≈ 56 × (−8.8 MeV)
B.E. ≈ −490 MeV

(c)(i) $^{226}_{88}Ra \rightarrow ^{222}_{86}Rn + ^{4}_{2}He$

Correct nucleon and proton numbers shown for the α particle and the daughter nucleus of radon Rn.

(ii) Mass defect $= \Delta m$

$\Delta m = (221.97 + 4.00) - 225.98$

$\Delta m = -0.01$ u

$\Delta m = -0.01 \times 1.66 \times 10^{-27} = -1.66 \times 10^{-29}$ kg

$\Delta E = \Delta mc^2$

$\Delta E = \Delta mc^2 = 1.66 \times 10^{-29} \times (3.00 \times 10^8)^2$

$\Delta E = 1.494 \times 10^{-12} \approx 1.49 \times 10^{-12}$ J

(iii) The energy is released as the kinetic energy of the α particle and the Rn nucleus.

Examiner's tip

Mass defect can means different things. In this question, the change in mass is identified as the energy released as kinetic energy of the by-products in the decay. According to Einstein's equation,

$$\Delta E = \Delta mc^2$$

*a **decrease** in mass means that energy is **released** in the decay. An increase in mass would imply that external energy is required for some event to occur.*

For an everyday event like walking, the change in our body mass is insignificant and we are therefore oblivious of it. For nuclear events, however, a tiny change in mass becomes extremely significant especially when there are a very large number of nuclei present.

(iv) The energy is released over a long period of time and therefore, the power of the source would be very small.

(8) Activity = rate of disintegration of nuclei.

Power = rate of energy released.

Therefore, power = activity × energy of each beta-particle.

The statement is incorrect because power also depends on the kinetic energy of each beta-particle released.

The activity is not constant over a period of time. As the nuclei decrease in number, so does the power.

Examiner's tip

The chances are that the examiners may also assess the candidate's 'Quality of Written Communication' in this question. It is therefore very important that extra care is given to sentence construction, spelling and grammar.

Mock Exam

Time: 1 hour 20 minutes   Maximum marks: 80

Instructions
Answer **all** questions in the spaces provided.
Show all steps in your working.
The marks allocated for each question are shown in brackets.
Any data required for a question is given where appropriate.

Grading
Boundary for A grade   64/80
Boundary for C grade   48/80

---

(1) **(a)** *Momentum* is a *vector* quantity.
Explain the meaning of the terms in italics.

....

.... **[3]**

(b) Explain what is meant by an **elastic** collision.

....

.... **[2]**

(c) Gas atoms in the Earth's atmosphere constantly collide with each other.
In one such collision, a hydrogen atom of mass 1.7×10^{-27} kg travelling at
420 ms^{-1}, makes a head-on collision with a stationary oxygen atom of mass
2.7×10^{-26} kg. After the impact, the oxygen atom moves with a speed v and
the hydrogen atom rebounds with a speed 370 ms^{-1}. This collision event is
shown in the diagram below.

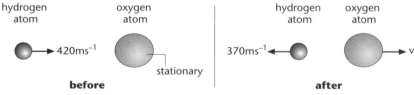

(i) State what is conserved in all collisions.

.... **[1]**

(ii) Calculate the magnitude of the velocity v of the oxygen atom.

....

....

.... **[3]**

[9 marks]

(2) **(a)** The diagram shows an ideal gas trapped inside a cylinder by means of a frictionless piston of surface area A. The pressure p exerted by the gas is kept constant as the piston is slowly moved out by a small distance Δx.

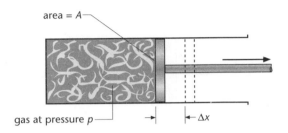

Show that the work W done by the expanding gas is given by
$$W = p\Delta V$$

....

.... **[3]**

(b) High pressure gas inside a spray-can has a volume of 400 cm³. The gas escapes into the atmosphere so that its final volume is six times that of the volume of the can.

(i) Calculate the work done by the expanding gas. Assume that the work is done against the constant pressure 1.0×10^5 Pa of the atmosphere.

....

.... **[3]**

(ii) Use the first law of thermodynamics to explain why the spray-can cools as the gas escapes from the can.

....

.... **[3]**

[9 marks]

(3) A cyclist is travelling on a banked circular track at a constant speed of 12 ms⁻. The cyclist is travelling in a horizontal circle of radius 40 m and is at right angles to the track as shown in the diagram.

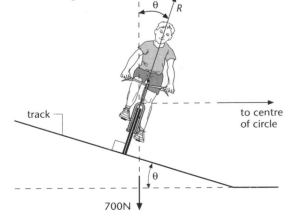

The combined weight of the cyclist and the bicycle is 700 N. The track exerts a contact force R on the cyclist. This force is at right angles to the track.

(a) For the combination of the cyclist and the bicycle, calculate
  (i) the centripetal acceleration,

  ....

  .... **[2]**

  (ii) the centripetal force.
      Data: $g = 9.8 \text{ Nkg}^{-1}$

  ....

  .... **[3]**

(b) Determine the angle θ made by the track with the horizontal.

  ....

  .... **[4]**

[9 marks]

(4) The universe is believed to have originated from a Big Bang some 1.8×10^{10} years ago. At the very early life of the universe, photons were constantly being converted into particles and vice versa.
When a particle and its anti-particle meet, they annihilate each other. The mass of the particles is converted into photons. One such event is the interaction between an electron and its anti-particle, the positron.

(a) Calculate the energy produced in the form of a photon when an electron and a positron annihilate each other.
Data: mass of electron $= 9.1 \times 10^{-31}$ kg
$c = 3.0 \times 10^8 \text{ ms}^{-1}$

  ....

  .... **[3]**

(b) Calculate the wavelength of the electromagnetic radiation after an interaction between an electron and a positron.
Data: $h = 6.63 \times 10^{-34}$ Js

  ....

  .... **[3]**

(c) State how the answer to **(b)** would change if a proton and an anti-proton were to be annihilated.

  .... **[1]**

[7 marks]

(5) The diagram shows the path of a proton travelling in a vacuum at 5.0×10^7 ms^{-1} as it enters and leaves a region of uniform magnetic field having a flux density of 1.4 T. The direction of the field is into the plane of the paper.

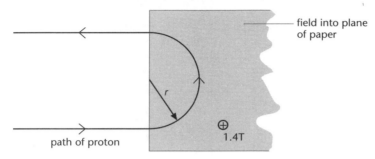

(a) Explain why the speed of the proton is not affected by the presence of the magnetic field.

....

.... [2]

(b) The proton describes a circular path in the region of the magnetic field. For the proton, calculate
   (i) its acceleration,
      Data: mass of proton = 1.7×10^{-27} kg
      e = 1.6×10^{-19} C

....

.... [3]

   (ii) the radius of the circular path in the field.

....

.... [2]

(c) Show that the time spent in the field by the proton is independent of its speed v.

....

.... [3]

[10 marks]

(6) (a) State Faraday's law of electromagnetic induction.

....

.... [1]

(b) The diagram shows a circular coil of area 3.5×10^{-4} m^2 and having 1 000 turns placed in a uniform magnetic field having a flux density B of 0.75 T. The coil is at right angles to the direction of the field.

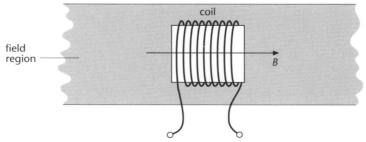

(i) State why there is no e.m.f. induced across the ends of the coil when the coil is moved at a constant speed in a direction parallel to the magnetic field lines.

.... **[1]**

(ii) Calculate the flux linkage for **each** turn of the coil.

....

.... **[3]**

(iii) The coil is removed from the magnetic field in a time of 20 ms. Calculate the magnitude of the mean e.m.f. induced across the ends of the coil.

....

.... **[3]**

(iv) State another way in which an e.m.f. may be induced across the ends of the coil.

.... **[1]**

[9 marks]

(7) (a) In 1961, Jonsson carried out experiments that provided further evidence that electrons were diffracted by very narrow slits. This is illustrated in the diagram below.

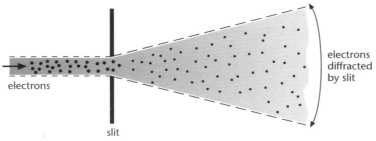

State what may be interpreted about the nature of electrons from such diffraction experiments.

.... **[1]**

(b) In a small television tube, electrons travel at a speed of 2.6×10^7 ms^{-1}. For an electron, calculate

(i) its momentum,

Data: mass of electron = 9.1×10^{-31} kg

....

.... **[2]**

(ii) the de-Broglie wavelength λ.

Data: $h = 6.63 \times 10^{-34}$ Js

....

.... **[2]**

(c) Explain why a person of mass 65 kg running at 9.0 ms^{-1} through an open door will not exhibit diffraction effects.

....

.... **[2]**

[7 marks]

(8) **(a)** The diagram shows two energy levels for a particular type of atom.

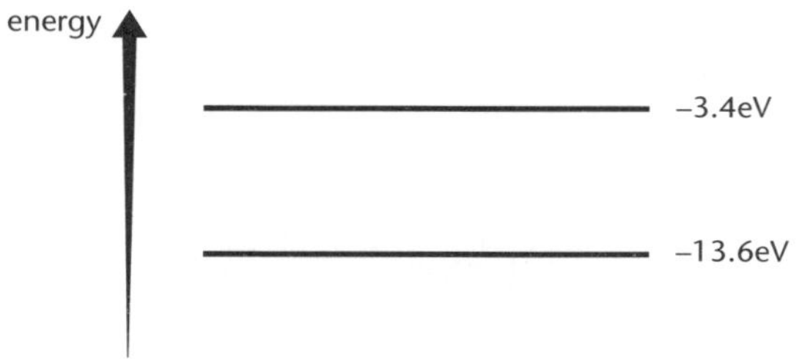

energy

———————————————— –3.4eV

———————————————— –13.6eV

(i) Explain why energy levels have negative values.

....

.... **[1]**

(ii) On the diagram indicate with an arrow the transition made by an electron responsible for emitting a photon from the atom.

[1]

(b) A laser produces a monochromatic light of wavelength 6.4×10^{-7} m.
The output power of the laser is 6.0 mW.

   (i) Calculate the energy of a single photon of light from the laser.
      Data: $h = 6.63 \times 10^{-34}$ Js
           $c = 3.0 \times 10^{8}$ ms^{-1}

....

....

.... **[3]**

   (ii) Hence determine the rate at which the photons are released
       from the laser.

....

.... **[2]**

(c) Without any further calculations, state and explain the effect on the
rate of photon release if the 6.0 mW laser in **(b)** were to produce high
frequency X-rays.

....

.... **[2]**

[9 marks]

(9) The escape velocity of an object is the minimum velocity that an object must have
in order to escape from the gravitational field of a planet or a star.

    (a) Show that the escape velocity v for an object is independent of its mass and is
given by

$$v = \sqrt{\frac{2GM}{R}}$$

where M is the mass of the planet and R is its radius.

....

....

....

.... **[3]**

(b) (i) Use the equation in **(a)** to calculate the escape velocity of protons on the surface of the Sun.
Data: mass of Sun = 2.0×10^{30} kg
radius of Sun = 6.7×10^{8} m.
$G = 6.67 \times 10^{-11}$ Nm^{2}kg^{-2}

....

....

.... **[2]**

(ii) Calculate the r.m.s. speed of protons on the surface of the Sun and comment on its value in relation to your answer to **(b)(i)**. You may assume protons behave like the molecules of an ideal gas.
Data: Sun's surface temperature ≈ 5 800 K
mass of proton = 1.7×10^{-27} kg
$k = 1.38 \times 10^{-23}$ JK^{-1}

....

....

.... **[3]**

(c) A black hole is a collapsed star that has an intense gravitational field from which even light cannot escape. Assuming the equation in **(a)** may be applied to light, calculate the maximum radius R of a black hole assuming it has a mass the same as that of our Sun.
Data: $c = 3.0 \times 10^{8}$ ms^{-1}

....

....

.... **[3]**

[11 marks]

Answers

(1) (a)   momentum = mv
where m is the mass of the object and v is its **velocity**.

A vector quantity has both magnitude **and** direction.

(b)   In an elastic collision both momentum and kinetic energy are conserved.

Examiner's tip

*In all collisions, energy is conserved. However, in your response you must make it clear that it is the **kinetic** energy that is conserved in an elastic collision. When the kinetic energy is converted into other forms, such as heat, sound etc., the collision is then referred to as an inelastic collision.*

(c)(i)   Momentum is conserved in all collisions.

(ii)   Total initial momentum = Total final momentum
$1.7 \times 10^{-27} \times 420 = (2.7 \times 10^{-26} \times v) + (1.7 \times 10^{-27} \times -370)$
$v = (7.14 \times 10^{-25} + 6.29 \times 10^{-25}) / 2.7 \times 10^{-26}$
$v = 49.7 \approx 50 \text{ ms}^{-1}$

Examiner's tip

Momentum is a vector quantity. After the collision, the hydrogen atom is moving in the opposite direction. Its momentum must therefore be assigned an opposite sign. To get the answer, you must first set up the equation in accordance with the principle of conservation of momentum and then proceed to solve it in terms of v.

(2) (a) $W = F\Delta x$
$p = \dfrac{F}{A}$
$\therefore\ W = (pA)\,\Delta x = p(A\Delta x)$
But $A\Delta x = \Delta V$, the change in volume
Hence $W = p\Delta V$

(b)(i) $\Delta V = 5 \times 400 \times 10^{-6}$
$\Delta V = 2.0 \times 10^{-3} \text{ m}^3$
$W = p\Delta V = 1.0 \times 10^5 \times 2.0 \times 10^{-3}$
$W = 200 \text{ J}$

Examiner's tip

*The initial volume of the gas is equal to the volume of the container. The final volume of the expanded gas is six times this initial volume. Therefore, the **change** in the volume of the gas must be five times the initial volume of the gas. There is another complication in this question. The volume is given in cm^3. Remember that*
$$1 \text{ cm}^3 = 10^{-6} \text{ m}^3.$$

(ii)   The first law of thermodynamics:
The change in internal energy = external heat supplied **to** the gas + work done **on** the gas.

The expanding gas does work against atmospheric pressure.
The internal energy of the gas provides the energy for this work to be done.
Since no external heating takes place, the internal energy must decrease.
The can therefore cools down.

Examiner's tip

There are many ways of writing the first law of thermodynamics. In order to avoid confusion with defining symbols, it is safer to write the law in words.

(3) (a)(i) $a = \dfrac{v^2}{r}$

$a = \dfrac{12^2}{40}$

$a = 3.6 \text{ ms}^{-2}$

(ii) $m = \dfrac{\text{weight}}{g}$

$m = \dfrac{700}{9.8} = 71.4 \text{ kg}$

$F = ma$

$F = 71.4 \times 3.6$

$F = 257 \approx 260 \text{ N}$

(b) Horizontally $\Rightarrow$ $R \sin \theta = 257$
Vertically $\Rightarrow$ $R \cos \theta = 700$

Therefore $\tan \theta = \dfrac{257}{700}$

$\theta = 20.2° \approx 20°$

Examiner's tip

The net force in the horizontal direction is responsible for the circular motion of the cyclist. There is no net force in the vertical direction. These two statements provide the necessary physics for this question. The question also requires a knowledge of the following piece of mathematics:

$$\tan \theta = \frac{\sin \theta}{\cos \theta}$$

(4) (a) $\Delta m = 2 \times 9.1 \times 10^{-31}$
$\Delta m = 1.82 \times 10^{-30} \text{ kg}$
$\Delta E = \Delta mc^2$ (Einstein's equation)
$\Delta E = 1.82 \times 10^{-30} \times (3.0 \times 10^8)^2$
$\Delta E = 1.64 \times 10^{-13} \approx 1.6 \times 10^{-13} \text{ J}$

(b) $E = hf$ and $c = f\lambda$
$\therefore E = \dfrac{hc}{\lambda}$

$\lambda = \dfrac{6.63 \times 10^{-34} \times 3.0 \times 10^8}{1.64 \times 10^{-13}}$

$\lambda = 1.21 \times 10^{-12} \approx 1.2 \times 10^{-12} \text{ m}$

Examiner's tip

The mass of the two particles is converted into a photon. The wavelength of the emitted electromagnetic radiation lies in the γ-ray region of the electromagnetic spectrum.

(c)   The wavelength would be shorter.
This is because the change in mass is greater, hence the
energy of the photon would be greater.

(5) (a)   The force on the proton is **perpendicular** to the velocity.
The speed is the same since no work is done by this force on the proton.

(b)(i) $F = Bqv$
$F = 1.4 \times 1.6 \times 10^{-19} \times 5.0 \times 10^{7}$
$F = 1.12 \times 10^{-11}$ N

$a = \dfrac{F}{m}$

$a = \dfrac{1.12 \times 10^{-11}}{1.7 \times 10^{-27}}$

$a = 6.59 \times 10^{15} \approx 6.6 \times 10^{15}$ ms^{-2}

(ii) $a = \dfrac{v^2}{r}$

$r = \dfrac{(5.0 \times 10^{7})^2}{6.59 \times 10^{15}}$

$r = 0.379 \approx 0.38$ m

(c) $F = ma$

$Bqv = m\left(\dfrac{v^2}{r}\right)$

$\left(\dfrac{v}{r}\right) = \dfrac{Bq}{m}$

$\text{time} = \dfrac{\text{distance}}{\text{speed}}$

$\text{time} = \dfrac{\pi r}{v}$

$\text{time} = \pi\left(\dfrac{m}{Bq}\right)$

The time spent in the field is independent of the speed v.

(6) (a)   The e.m.f. induced in a circuit $\propto$ rate of change of flux linkage.

(b)(i) There is no change in flux linking the coil.

(ii) flux linkage = NBA

For one turn, flux linkage = ϕ ($N = 1$)

$\phi = BA$

$\phi = 0.75 \times 3.5 \times 10^{-4}$

$\phi = 2.63 \times 10^{-4} \approx 2.6 \times 10^{-4}$ Wb

Examiner's tip

The flux linking each turn is given by

$$\phi = BA \cos \theta$$

where θ is the angle between the normal of the coil and the magnetic field. Since the magnetic field is normal to the plane of the coil, the angle θ is equal to 0° and therefore the cos θ factor is equal to unity.

(iii) e.m.f. = − rate of change of flux linkage

change in flux linkage = $1000 \times 2.63 \times 10^{-4}$ Wb

$$\text{e.m.f.} = \frac{1000 \times 2.63 \times 10^{-4}}{20 \times 10^{-3}}$$

e.m.f. = $13.2 \approx 13$ V

(iv) Rotate the coil in the magnetic field.

Examiner's tip

The flux linking each turn of the coil is given by

$$\phi = BA \cos \theta$$

where θ is the angle between the normal of the coil and the magnetic field. By changing the angle θ, the flux linking each turn may be altered with respect to time and therefore an e.m.f. induced across the ends of the coil.

(7) (a) **Moving** electrons behave like **waves**.

Examiner's tip

*The waves associated with moving particles are **not** electromagnetic in nature. They have their own peculiar characteristics. The waves are referred to as either 'de-Broglie waves' or 'matter waves'.*

(b)(i) $p = mv$

$p = 9.1 \times 10^{-31} \times 2.6 \times 10^7$

$p = 2.37 \times 10^{-23} \approx 2.4 \times 10^{-23}$ kgms^{-1}

(ii) $\lambda = \dfrac{h}{p}$

$$\lambda = \frac{6.63 \times 10^{-34}}{2.37 \times 10^{-23}}$$

$\lambda = 2.80 \times 10^{-11} \approx 2.8 \times 10^{-11}$ m

(c) The wavelength of the person is too small (compared with the width of the door).

$$\lambda = \frac{6.63 \times 10^{-34}}{(65 \times 9.0)} \sim 10^{-36} \text{ m}$$

Examiner's tip

For diffraction effects to be prominent, the wavelength λ must be comparable to size of the 'gap'. For a person to show diffraction effects, the gap size ought to be about 10^{-36} m, which is an impossibility. Electrons may be diffracted by matter because their de-Broglie wavelength can be comparable to the separation between the atoms.

(8) (a)(i) Negatively charged electrons are bound to the positive nucleus. External energy is required to either excite the atom or to remove the electron completely from the influence of the positive nucleus.

(ii) Transition shown from the –3.4 eV energy level to the –13.6 eV energy level.

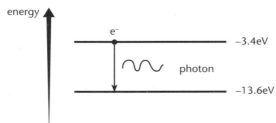

Examiner's tip

*An electron making the above transition is **losing** energy. In terms of energy, the atom is much more stable after this event. Since energy must be conserved, the energy of the electron is transformed into a **photon** of electromagnetic radiation.*

(b)(i) $f = \dfrac{c}{\lambda}$

$f = \dfrac{3.0 \times 10^{8}}{6.4 \times 10^{-7}}$

$f = 4.69 \times 10^{14} \text{ Hz}$

$E = hf$

$E = 6.63 \times 10^{-34} \times 4.69 \times 10^{14}$

$E = 3.11 \times 10^{-19} \approx 3.1 \times 10^{-19} \text{ J}$

(ii) Rate of photon release = power of laser / energy of each photon

$N = \dfrac{6.0 \times 10^{-3}}{3.11 \times 10^{-19}}$

$N = 1.93 \times 10^{16} \approx 1.9 \times 10^{16} \text{ s}^{-1}$

(c) The energy of a single X-ray photon is greater. Hence for the **same** power, there are fewer photons released in a given time.

(9) (a) $P.E. = \dfrac{-GMm}{r}$

$K.E. = \frac{1}{2}mv^2$

The kinetic energy at the surface must be sufficient to overcome the gravitational potential energy, therefore

$K.E. = P.E.$

$\frac{1}{2}mv^2 = \dfrac{GMm}{r}$

$v = \sqrt{\dfrac{2GM}{R}}$

(b)(i) $v = \sqrt{\dfrac{2GM}{R}}$

$v = \sqrt{\left(\dfrac{2 \times 6.67 \times 10^{-11} \times 2.0 \times 10^{30}}{6.7 \times 10^8}\right)}$

$v = 6.31 \times 10^5 \text{ ms}^{-1} \approx 6.3 \times 10^5 \text{ ms}^{-1}$

(ii) $\frac{1}{2}m\langle c^2 \rangle = \frac{3}{2}kT$

$\langle c^2 \rangle = \dfrac{3kT}{m} = \dfrac{3 \times 1.38 \times 10^{-23} \times 5800}{1.7 \times 10^{-27}}$

$\langle c^2 \rangle = 1.41 \times 10^8 \text{ m}^2\text{s}^{-2}$

r.m.s. speed $= \sqrt{1.41 \times 10^8} = 1.19 \times 10^4 \text{ ms}^{-1}$

r.m.s. speed $\approx 1.2 \times 10^4 \text{ ms}^{-1}$

The r.m.s. speed is far less than the escape velocity, therefore the protons remain on the surface of the Sun.

Examiner's tip

The r.m.s. speed of the protons represents an 'average' value. Some protons, at the surface temperature of 5 800 K, will be travelling fast enough to overcome the attractive force of the Sun. Hence, some protons do escape.

(c) $R = \dfrac{2GM}{v^2}$

$R = \dfrac{2 \times 6.67 \times 10^{-11} \times 2.0 \times 10^{30}}{(3.0 \times 10^8)^2}$

$R = 2.96 \text{ km} \approx 3.0 \text{ km}$